ROBOTICS/
ARTIFICIAL INTELLIGENCE/
PRODUCTIVITY:
U.S. — Japan
Concomitant Coalitions

ROBOTICS/ ARTIFICIAL INTELLIGENCE/ PRODUCTIVITY: U.S.—Japan Concomitant Coalitions

BOOK I: Productivity Promises of Robotics and Artificial
Intelligence
Preface
Introduction
BOOK II: Concomitant Coalitions in Cars and Computers

George K. Chacko
UNIVERSITY OF SOUTHERN CALIFORNIA
Institute of Safety and Systems Management

PETROCELLI BOOKS
Princeton, New Jersey

Dedicated Affectionately To Our Son
RAJAH
whose intelligence is not artificial

Library of Congress Cataloging-in-Publication Data

Chacko, George Kuttickal, 1930-
 Robotics/artificial intelligence/productivity.

 1. Flexible manufacturing systems—United States. 2. Flexible manufacturing systems—
Japan. 3. Robotics—United States. 4. Robotics—Japan. 5. Computer industry—United
States—Automation. 6. Computer industry—Japan—Automation. 7. Automobile industry
and trade—United States—Automation. 8. Automobile industry and trade—Japan—
Automation. I. Title.
HD9725.C48   1986   658.5'14   86-12177
ISBN 0-89433-228-7

Contents

v

Robotics/Artificial Intelligence/Productivity

Book Two
Concomitant Coalitions in Cars and Computers

ABOUT THE AUTHOR

"It pleases me very much to see that the theory of games has found such an able expositor, and I hope you will be among those who will expand the work originally," wrote Professor Oskar Morgenstern, co-inventor with Professor John von Neumann, of The Theory of Games, to 19-year-old Chacko, completing his work for his Master's Degree in India. In 1960, he originated the concept of Concomitant Coalitions while working on a production and distribution problem, first published in *Operations Research* in 1961.

Author, editor and/or contributor of 33 books, with 176 professional publications and presentations to his credit, Dr. Chacko has followed the developments in computers since 1950 when, as a fellow at the Indian Statistical Institute, he was exposed to the Hollerith Machine. His invention of a highly accurate, computer-based, numerical forecasting method using very few data points, MESGRO (*Mod*ified *E*xponential *S*moothing - *GRO*wth Stage), won him a U.S. National Science Foundation competitive award for International Scientific Lectures in 1982.

Manager of Operations Research Department of Hughes Semiconductor Division and Senior Staff Scientist with TRW Systems before joining University of Southern California in 1970, where he is Professor of Systems Science, Dr. Chacko was Senior Fulbright Professor of Management Science at National Chengchi University, Taipei in 1983. He returned there as Senior Fulbright Research Professor in 1984, to work on a Taiwan National Science Council Research Project entitled: "Concomitant Coalitions in Trade and Technology."

BOOKS BY GEORGE K. CHACKO

AS AUTHOR

The Mar Thoma Syrian Liturgy, A Translation into English, 1956.
The Mar Thoma Syrian Church Order of Holy Matrimony, A Translation into English, 1957.
India—Toward an Understanding, 1959.
International Trade Aspects of Indian Burlap—An Econometric Study, 1961.
Today's Information for Tomorrow's Products—An Operations Research Approach, 1966.
Studies for Public Men, 1969.
Applied Statistics in Decision-Making, 1971.
Computer-Aided Decision-Making, 1972.
Technological Forecontrol—Prospects, Problems, and Policy, 1975.

Applied Operations Research/Systems Analysis in Hierarchical Decision-Making, 1976:
Volume I—Systems Approach to Public and Private Sector Problems, 1976.
Volume II—Operations Research Approach to Problem Formulation and Solution, 1976.
Management Information Systems, 1979.
Life Abundant Day by Day, 1985.
Interceding with the Infinite, 1985.

Robotics/Artificial Intelligence/Productivity

Robotics/Artificial Intelligence/Productivity:
  U.S.—Japan Concomitant Coalitions, 1985.

AS EDITOR, CONTRIBUTOR

Management Sciences: Models and Techniques (contributor), 1961.

Long-Range Forecasting Methodology (contributor), 1968.

Reducing the Cost of Space Transportation (editor), 1969.

The Recognition of Systems in Health Services (editor and contributor), 1969.

Systems Technology Applied to Social and Community Problems (contributor), 1969.

Planning Challenges of the '70s in the Public Domain (co-editor), 1970.

Congressional Recognition of Goddard Rocket and Space Museum (contributor), 1970.

Systems Approach to Environmental Pollution (editor and contributor), 1971.

Hope for the Cities—Systems Approach to Human Needs (contributor), 1971.

The Use of Modern Management Methods in the Public Administration of Developing Countries (contributor), 1972.

Alternative Approaches to the National Delivery of Health Care (editor), 1972.

Systems Approach to Strokes and Heart Diseases (co-editor), 1978.

Health Handbook (editor), 1979:

Vol. I—Environmental Management for Improved Health Services—U.S., U.K., Europe, 1979.

Vol. II—National Organization of Health Services—U.S., USSR, China, Europe, 1979.

Vol. III—Computer Augmentation of Health Service Operations: Diagnostics to Decision-Making—U.S., Europe, Africa, Asia, Australia, 1979.

Vol. IV—Educational Innovation in Health Services—U.S., Europe, Middle East, Africa, 1979.

Vol. V—Health Indicators and Health Services Utilization—U.S., Europe, 1979.

Preface

This work presents a unified treatment of the promises of productivity of the many emerging, expanding, and exploding technologies under the rubrics of robotics and artificial intelligence, on the one hand, and the competitive cooperation between the U.S. and Japan to secure and share the advances and applications in these fields, on the other. The focus on productivity makes the treatment highly practical from the point of view of both technology and trade.

The volume is divided into two books. Book I is entitled: "Productivity Promises of Robotics and Artificial Intelligence," and Book II, "Concomitant Coalitions in Cars and Computers."

Part I: "Productivity Perspectives" opens with a study in Chapter 1: "Robotize or Retrogress?" of the continuing Industrial Revolution providing the setting for the coming Robotics Revolution. While the enlarging territory of the eighteenth century British Empire stimulated the technology of mechanization, today in a reverse process, the technology of microminiaturization stimulates the enlarging of international market territory, vigorously pursued by Japan and others. The sustained success of the Industrial Revolution came from increased productivity, which is likely to be the issue in the Robotics Revolution at home and abroad.

Chapter 2: "The Sources and Signals of Declining U.S. Productivity" identifies the sources and causes of productivity decline in the U.S. In contrast, "Growth agreed upon and growth pursued" characterize Japanese production and enhances productivity. Chapter 3: "The Product, Process, and Premises of Rising Japanese Productivity" groups the factors of Japanese productivity according to the interrelated and integral elements of a triad—institutional, investmental, and inventional/innovational—, which explains the administrative guidance, the beneficial guidance to banks, and the special stimulus to companies to invest in research and development.

Part II: "Improving Productivity through Integrated Manufacturing" discusses the control *of* the metal in Chapter 4: "Robots, CAD/CAM, and other Automation Technologies in Integrated Manufacturing," while highly parallel computing *à la* the Fifth Generation Computer Systems could be considered control *in* the metal. Recent developments in four potential areas of savings in manufacturing: assem-

bly, feeding, programmable assembly, and design, are discussed in Chapter 4, as well as the American, Japanese, and West German experience in savings in manufacture. At the heart of successful automation of manufacturing is the control exercised through communication. Chapter 5: "Implementing Integrated Controls—Hardware and Software" discusses robot dynamics and control. The imbedding of controls in the machine attempts to make hardware progressively take over the software functions.

Productivity promises of robotics—the *metal*—depend on the progress in artificial intelligence—the *mental*, which is the subject of Part III: "Improving Productivity through Artificial Intelligence." Impressive as the reductions in cost of manufacturing are, the maximum contribution by robotics substitution is only 4.8 percent of the selling price, while marketing, sales and general administration and engineering together represent 40 percent of the selling price. Chapter 6: "Reasoning Robots for Efficient Enterprises" examines the potential of Artificial Intelligence in increasing productivity in non-manufacturing areas. Chapter 7: "Knowledge-processing for Better Decision-making" focuses on humanlike capabilities with special reference to major expert systems and Distributed Problem-solving.

To realize the advances in technology, and to reap the benefits of applications in trade, the U.S. and Japan will necessarily compete in some areas and co-operate in some other, as evidenced in GM-Toyota competing for the world market for special subcompacts that are actually produced cooperatively. This association of one party both with and against other party(ies) in the same game or activity is called Concomitant Coalition (CONCOL).

Book II: "Concomitant Coalitions in Cars and Computers" discusses CONCOLs in trade and technology, focusing on Japan and the U.S. It comprises two parts. Part IV: "Concomitant Coalitions (CONCOLs) in Japan and in the United States" opens Chapter 8: "CONCOLs in Trade and Technology" with a new concept, Adaptive Automatics, subsuming robotics and artificial intelligence, illustrated elegantly in nature. To realize CONCOLs, the players have first to be identified, Chapter 9: "CONCOL Players promoting Productivity in Japan" identifying the players in eight institutional, nine investmental, and seven inventional/innovational elements. While in Japan, the alter ego in the institutional, investmental, and inventional/innovational elements is more often than not successful in having its long-term planning prevail over the anti-ego's short-term profits view, it is not so in the United States, as seen in Chapter 10: "CONCOL Players in Robotics in the United States."

After identifying the CONCOL Players in Japan and in the U.S. in Part IV, Part V: "Concomitant Coalitions (CONCOLs) between the United States and Japan" turns to the CONCOL Process in Chapter 11: "The Context of the CONCOL Process." A quantitative initial desirable outcome is identified with respect to the seven macroelements, four minielements, and three microelements as providing the context of the CONCOL process with respect to high-technology and high-technology products trade in general, and Fifth Generation (5G) Computer and as-

sociated products, in particular. The 5G is not a computer; it is a revolution. Chapter 12: "Cars and Computers: Keys to U.S.—Japan CONCOLs" focuses on 5G-based products although the prototype 5G itself is not scheduled until 1990, and develops a detailed illustration of the manufacture and marketing of cars under 5G technology in five markets. While the least visible at the moment, 5G technology potentials in management are likely to exert a profound effect on how the world thinks and trades, plays and lives in the coming decades.

Introduction

"I've got a question in my mind," says [Roger B.] Smith, [chairman of the board of General Motors]. "Do we need to do anything more in that area? . . . There's enough horses in that field out there for nine races, and I don't just want to win one race. So maybe we have to go out there and get a couple of more horses. I don't think I know right now that I want to say, Okay, I made my bet on GM-Fanuc. It could run in the sixth at the Belmont, and that's all I've ever got to do to be the biggest horseman in the country."[1]

The *Washington Post* interviews with Mr. Smith and others of General Motors clearly indicate that the "field" is not by any means limited to the manufacture and sale of factory robots and robotics. If anything, the "field" is limited only by imagination, ingenuity, and investment. Since investment on the scale of General Motors, as well as of others on a scale less grand, demands committing resources today for results tomorrow, a profile of the possible is indispensable.

Robotics/Artificial Intelligence/Productivity examines the productivity promises of the many emerging, expanding, and exploding technologies under the rubrics of robotics and artificial intelligence. Progress in robotics, the *metal*, is dependent upon progress in artificial intelligence, the *mental*. Book One, "Productivity Promises of Robotics and Artificial Intelligence," comprises three parts: "Productivity Perspective," "Improving Productivity Through Integrated Manufacturing," and "Improving Productivity Through Artificial Intelligence."

Neither advances in nor application of these productivity promises is the monopoly of the United States. In fact, Japan initiated the Fifth-Generation Computer Systems Project as a national project in fiscal 1982, in April, setting up the Institute for New Generation Computer Technology (ICOT) as the principal organization for executing the project. Its goal is to produce a prototype Knowledge Processing System (KPS) by 1990.

Technology is tested in trade. Whether it is the Fifth-Generation Computer or the First-Generation Robot, productivity promises can be realized only to the ex-

---

1. Michael Schrage and Warren Brown, "GM: Chrome to Computers," Part 2, *The Washington Post*, 5 July 1984, A4.

tent that the new products and processes command the market. And the market is increasingly international.

In 1982, the U.S.-transpacific trade at \$121.2 billion exceeded the U.S.-transatlantic trade at \$115.8 billion. Fueling the growth of the transpacific trade is high technology. Importing high technology from the U.S. and Japan in order to export high-tech products to these countries places Taiwan, Korea, Hong Kong, Singapore, and other similar trade partners in opposing roles: Taiwan simultaneously *with* the U.S., paying more for high-tech, and *against* the U.S., demanding more for exported high-tech products. This association of one party both with and against other party(ies) in the same game is called *Concomitant Coalition* (CONCOL).

Book Two, "Concomitant Coalitions in Cars and Computers," discusses CONCOLs in trade and technology, focusing on Japan and the U.S. It comprises two parts: "Concomitant Coalitions (CONCOLs) in Japan and in the United States" and "Concomitant Coalitions (CONCOLs) Between the United States and Japan."

The two books—"Productivity Promises" and "Concomitant Coalitions"—are integral to each other, because successful concomitant coalitions among trading partners depend on an accurate assessment of high-tech productivity promises. When Mr. Smith asks if he should acquire additional capabilities in robotics and artificial intelligence, he is asking for a profile of advances and applications in an exploding field. No one would want to say about any application, "This is it. Bet on it." No one would want to be ignorant of any potential advance of technology that transform our day-to-day world either, when deciding whether or not to invest, how much, and when.

Japan and the United States differ on the very definition of industrial robots, but irrespective of the definition, Japan is undeniably the front-runner in industrial robots. Is robotics really necessary for American manufacturing? Yes, according to the National Academy of Sciences, whose Committee on Computer-Aided Manufacturing points out that manufacturing is responsible for two-thirds of wealth-creation. By reducing the cost of manufacture, computer-integrated manufacturing offers very great potential to reduce the cost of generating wealth. Focusing on specific American, Japanese, and West German experiences with integrated manufacturing, chapter 4 discusses recent developments in four potential areas of savings in manufacturing: *assembly, feeding, programmable assembly,* and *design.*

Robot manipulation, including robot vision, could be considered control *of* the metal, while highly parallel computing *à la* the Fifth-Generation Computing System could be considered control *in* the metal. Robot dynamics and control are accomplished through commands which the robot "understands," so, through a comparative study of high-level robot languages, it is essential to identify the most programmable language. Since one of the most effective ways that a robot can interact with external devices and sensors is through robot vision, chapter 5 reviews

current developments in the process of computer vision for robot control under the following headings: "Sensing," "Segmentation," "Description," "Recognition," and "Interpretation." Imbedding controls in the machine, as is done in machine vision, attempts to make hardware progressively take over software functions, so chapter 5 also ranks various chip architectures with respect to the system parameters.

While robotics in manufacturing has made possible impressive reductions in cost, as illustrated by a 30 percent reduction in the work force at Messerschmidt in West Germany, the handling of 97 percent of the work at Nissan's Zama auto plant in Japan, a 25 percent reduction in the work-in-process inventory at the John Deere Company in Waterloo, Iowa, etc., all of manufacturing represents only 40 percent of the selling price. Direct labor costs represent only 12 percent of the manufacturing cost, leaving as the target for cost reduction by robotics substitution only 4.8 percent of the selling price. In contrast to manufacturing, the traditional target area for cost reduction through robotics, two other areas together account for 40 percent of the selling price: marketing, sales, and general administration (25%), and engineering (15%). Productivity increases in marketing require that achievable goals be adaptively set (*goal-setting*)—instead of merely meeting goals (*goal-meeting*). Chapter 6 asks, "Will artificial intelligence (AI) be able to offer goal-setting capabilities?" AI methods include heuristic search, generate-and-test, theorem-proving, backward-chaining, forward-chaining, etc. Imbedding one or more of these methods to represent human problem-solving has been attempted in many algorithms. Chapter 6 describes several of these. The AI learning process in virtually all instances is *iterative*, while in real life, the process is almost always *inter*active.

The performance record of the iterative process encouraged the Defense Advanced Research Projects Agency (DARPA) to commit in 1983 close to $1 billion to develop by 1989 machines with humanlike capabilities, allowing them to see, reason, plan, and even supervise the actions of military systems in the field. The most celebrated among the computer systems that mimick human problem-solving are PROSPECTOR, which identified $100 million worth of molybdenum deposit, and CADUCEUS, which accurately diagnoses diseases from symptoms. Chapter 7 develops a new definition of expert systems and discusses the major ones. Almost all known expert systems, however, employ *iterative* processes, while important, real-life problems affecting productivity demand a problem-solving capability that is *interactive*.

Even while productivity promises are but a gleam in the eye of the scientist and the engineer, the entrepreneur has to identify potential products. Even more, he or she has to aggressively stake out a specific part of the possible market for the potential product. In 1972, Japan had not even made the first commercial microelectronic chip when they announced their national plans, and within a decade half the world's market for advanced memory chips was Japanese. No wonder General Motor's Roger Smith says that he can *create* the market: "I honestly

believe that if we get the right products, we can literally create the market. Now that's a wild statement to make in some regards . . . but I think if we come up with the electronics the way I see them coming with robots and other electronic things I see coming down the road, I believe that we can make the product so effective we can create the market."[2]

Competitive cooperation is the smart way to approach the increasingly international marketplace for high-tech products, be they cars or computers. The GM-Toyota agreement made it possible to reopen GM's shut-down $1.5 billion Fremont plant with $1.5 billion worth of robotics from Toyota to convert it into an automated factory. This is a recent example of the two largest auto manufacturers in the world *competing* for the world market for special subcompacts that are actually produced *cooperatively*. Again, the GM-Fanuc agreement, which created GMF Robotics, embodies the simultaneous *cooperation in manufacturing* robotics and *competition in marketing* robotics.

Competitive cooperation is the essence of Concomitant Coalition (CONCOL). To set the stage for CONCOLs in trade and technology, chapter 8 introduces robotics and artificial intelligence (respectively, the *metal* and the *mental*) as part of *Adaptive Automatics*. It envisages a First-Generation ADAPTOMATON as a self-discovering and self-designing system that transforms inputs into outputs progressively more efficiently in order to accomplish designed and/or discovered objectives. How should countries which import high technology in order to export high-tech products that embody the imported technology develop a viable strategy for competitive cooperation?

CONCOL is germane not only to *inter*national associations, but also to *intra*national associations. From the point of view of CONCOL, chapter 9 considers the *institutional*, *investmental*, and *inventional/innovational* factors of Japanese productivity. It identifies the CONCOL players in eight institutional, nine investmental, and seven inventional/innovational elements, and it reconciles or harmonizes the opposing interests through the CONCOL bargaining process.

Chapter 10 shows that balancing the opposing concerns of labor, capital, and technology *within* the U.S. is critical to success in balancing the opposing concerns of high-technology imports followed by exports of high-tech products in CONCOLS *between* the U.S. and trade partners such as Japan, Taiwan, Korea. Three critical elements of the process of CONCOL are: *areas of bargaining* (Why?), *measurement of outcome* (What?), and *changes of bargaining* (How?). Chapter 11 identifies a quantitative initial desirable outcome with respect to seven macroelements, four minielements, and three microelements. These provide the context of the CONCOL process with respect to high-technology and high-tech products trade in general, and the Fifth-Generation Computer and associated products in particular.

2. *Ibid.*

Introduction

The Fifth Generation is not a computer; it is a revolution.

To illuminate the revolutionary potential of the Fifth Generation, chapter 12 briefly reviews computer technology development in terms of three phases: *from data processing to information processing, from information processing to decision support systems*, and *from decision support systems to knowledge-processing systems*. The Fifth Generation will be a knowledge-processing machine by virtue of three interrelated capabilities. It will understand speech, graphics, and natural language; it will choose the most appropriate knowledge-base; and it will solve problems. While progress is being made toward realizing these capabilities, attention must be focused on products based on the Fifth Generation. The revolutionary potential of Fifth-Generation technology lies in the areas of manufacturing, marketing, and management.

In manufacturing, the Fifth Generation holds promise in improved, innovated, and individualized quality. These capabilities will impact the marketing of Fifth-Generation products. The impact can be specified in terms of the ratio of Performance Specific (PS) products to Performance General (PG) products, the former being custom-manufacture items, and the latter items of mass manufacture. When and where should PS and PG products be made and sold? There are five critical factors in the optimum selection of the product mix and manufacturing locations. Chapter 12 gives a detailed illustration of the manufacture and marketing of cars using Fifth-Generation technology in five markets. The third area holding revolutionary potential from Fifth-Generation technology is management. While the least visible at the moment, this potential is likely to exert a profound effect on the way the world thinks and trades, plays and lives in the coming decades.

This work is designed for everyone interested in the coming Robotics Revolution, a revolution propelled by the twin forces of technology and territory, a new and enormous explosion in the continuing Industrial Revolution. It is designed for use in graduate and advanced undergraduate courses in a variety of disciplines and schools—business, computers, systems, economics, technology, etc. It should also be useful to practitioners in industry who need to translate the esoteric developments in emerging fields into highly practical decisions of business strategy relating to manufacturing and nonmanufacturing industries, as well as military and space industries. Written in a readable style, making no assumptions of prior knowledge of computers, robotics, or artificial intelligence, this work hopefully contributes to a reasoned awareness of what is perhaps the single most significant development of our times.

BOOK ONE
Productivity Promises of Robotics and Artificial Intelligence

PART I
Productivity Perspectives

OVERVIEW: The twin forces of technology and territory, whose turbulence spawned the Industrial Revolution some two hundred years ago, today thrust forth the Robotics Revolution. While the enlarging territory of the eighteenth century British Empire stimulated the technology of mechanization, today in a reverse process, the technology of microminiaturization stimulates the enlarging of international market territory, vigorously pursued by Japan and others. With a fifth or more of American-owned cars Japanese-made, to catch up, the U.S. has to robotize, failure to do which will make the U.S. retrogress.

Are the stated reasons for resistance to robotics in the U.S. supported by past experience with the Industrial Revolution? Seeing that one machine could replace thirty men, the workers completely destroyed mills valued at £ 10,000 in one day. They had to give up their self-set pace of work, go to the factory, and be subject to rigorous discipline. Above all, the greed and brutality of early capitalism fueled the opposition of workers to mechanization. In the meantime, expanding markets at home and abroad required far greater capability to produce goods and services than what the guilds were able to provide. Unless mechanized, England would lose to other countries the enlarging market.

Today, Japan recognizes quite well the imperatives of international markets, which she pursues with all the advantages of the Robotics Revolution. There is, however, a price to pay. In spite of a fourth or more of Japanese employees working for the same employer all their life, Professor Hasegawa anticipates unemployment consequences of robotics revolution for both unskilled and skilled labor. In the United States, the U.S. Department of Labor estimates that robotics could displace 35 percent of manufacturing employees in the next decade or two.

Despite devastating unemployment, the major elements that changed the workers' attitude toward mechanization in the Industrial Revolution were the market which generated enormous demand for the products, the real threat of national ruin if improved productivity through mechanization were not adopted, the acceptance by the judges and the lawmakers of the inevitability of hardships, and the increased earnings emerging as a clear possibility. The sustained success of the Industrial Revolution came from increased productivity which is likely to be the issue in the Robotics Revolution at home and abroad.

1

Robotize or Retrogress?

0.0 From the Continuing Industrial Revolution to the Coming Robotics Revolution

The Industrial Revolution, which originated in England in the second half of the eighteenth century, is a *continuing revolution*. It reaches out to the ends of the earth, and it is rewriting contemporary history. The current phase is especially visible from the perspective of economic productivity, productivity that stems from new technology. The resulting cornucopia of goods and services pours out to territories far beyond the centers of production. In fact, the mutual reinforcement of technology and territory, whereby the demands of the one stimulate the supply of the other, is a recurring theme of the Industrial Revolution that has immediate parallels in the *coming revolution*, the Robotics Revolution.

The twin forces of technology and territory, whose turbulence spawned the Industrial Revolution some two hundred years ago, are emerging today in altered form, but with no less impetus to change, thrusting forth the Robotics Revolution. If the technology of mechanization in the form of the spinning jenny and the steam engine ushered in the Industrial Revolution, today the technology of microminiaturization in the form of transistors and microchips heralds the Robotics Revolution.

In eighteenth century England, the demands of enlarging territory stimulated the supply as well as the use of technology. Inventions by themselves do not create innovations, much less revolutions. Of critical importance to the Industrial Revolution was the emergence of "projectors," entrepeneurs who specialized not only in meeting demands, as the guild masters had done, but also, and more importantly, in creating demands:

> Colonial and other overseas trade gave British industry a foreign market unrivaled in the eighteenth century. Commerce with the English settlements in North America employed 1,078 vessels and 29,000 seamen. London, Bristol, Liverpool, and Glasgow flourished as chief ports for this Atlantic trade. . . .

Robotics/Artificial Intelligence/Productivity

> The guilds were not competent to meet the demands of expanding markets at home and abroad. They had been instituted chiefly to supply the needs of a municipality and its environs; they were shackled by old regulations that discouraged invention, competition, and enterprise; they were not equipped to produce raw materials from distant sources, or to acquire capital for enlarged production, or to calculate, obtain or fill orders from abroad. Gradually the guild master was replaced by "projectors" (entrepreneurs) who knew how to raise money, to anticipate or create demand, to secure raw materials, and to organize machines and men to produce for markets in every quarter of the globe.[1]

Could there be in Durant's description of the *continuing revolution* a clue to the conditions that facilitate the *coming revolution*?
- "not competent to meet the demands of expanding markets at home and abroad"
- "shackled by old regulations that discouraged invention, competition, and enteprise"
- "not equipped to acquire capital for enlarged production"
- "[not equipped] to calculate, obtain, or fill orders from abroad"

In short, could the coming revolution in robotics be stymied by the paucity of "projectors," of entrepreneurs?

The entrepreneurs of Industrial Revolution had an ardent advocate in Adam Smith, who, in his 1776 book, *The Inquiry into the Nature and Causes of the Wealth of Nations*, advocated free enterprise, which would thrive best when least interfered with by the government. The "Invisible Hand" of individual self-interest of the entrepreneurs, would lead to the collective betterment of society as a whole.

Turning from the Industrial Revolution and "the passage of the [British] economy from regulated guilds and home industry to a regime of capital investment and free enterprise,"[2] to the coming Robotics Revolution, the mutual reinforcement of technology and territory suggests itself as a viable theme, though not necessarily in the Industrial Revolution sequence; more likely the reverse. In the Robotics Revolution, it is technology that appears to seek territory.

Even in its infancy, the Robotics Revolution has stimulated another island-nation to launch an aggressive search for economic territories in the international marketplace. With the Industrial Revolution, the "foreign market unrivaled in the

---

1. Will and Ariel Durant. *Rousseau and Revolution*, Vol. X of *The Story of Civilization* (New York: Simon and Schuster, 1967), 669–70.
2. *Ibid.*, 680.

eighteenth century" hastened the application of technology in England "to meet the demands of expanding markets at home and abroad." With the Robotics Revolution in the 1980s, no such expanding markets already exist in the sense of an assured selling place for robotics wares. Markets have to be found. They have to be diverted, as they were, for instance, when demand for automobiles changed from American to Japanese make. When Japan can manufacture and export to the United States cars of comparable quality for about $700 less, that suggests a triumph, in part, of robotics technology, which, in turn, seeks to conquer territories.

The first industrial "playback robot" was made in the United States in 1962. It was introduced into Japan in 1967, and in the following decade and a half, Japan surged ahead to become the leading manufacturer of robots, while the United States lagged behind, not only in the number of robots manufactured, but in the number of robots used in industry as well. Will this trend continue? Or will the United States catch up and surpass?

The stakes are quite high. The dramatic shrinkage in the domestic market for U.S.-made automobiles is creating painful reverberations across the entire spectrum of U.S. industries, forcing a decisive response. If a fifth or more of American-owned cars are Japanese-made today, can that domestic market be recaptured at all? For such a dramatic reversal, the American car owner has to be shown conclusively that Detroit can indeed match and exceed the quality of the Japanese-make automobile, and that it can sell it competitively. This certainly means that the U.S. has to appropriate even better the fruits of the infant Robotics Revolution. To robotize or not to robotize is *not* the question, but how to robotize, how well, and how fast. If to catch up, the U.S. has to robotize, then failure to robotize would make the U.S. retrogress. So the question is: to robotize or to retrogress?

The very facts that promote the Robotics Revolution in Japan, which built its revolution on the first American industrial robot, may not let it thrive in the U.S. and may even serve to thwart it. Are the stated reasons for the present resistance to robotics in the U.S. today supported by past experience with a major industrial shift? Two hundred years ago, were the worst fears of the workers resisting the Industrial Revolution in its early days realized? How did the early resistance, which exploded into repeated riots, give way to enthusiastic embrace?

0.0 Workers of This Industry Unite!
You Have Nothing to Lose But Your Jobs!

The first legislative act passed to put down the riots against machinery in the second half of the eigtheenth century, dated 1769, made the penalty for willful destruction of any building containing machinery the same for arson,—death. Nevertheless, resistance to the coming machinery continued. Paul Mantoux, in *The Industrial Revolution in the Eighteenth Century*, writes:

7

Robotics/Artificial Intelligence/Productivity

This drastic law did not, however, prevent the recurrence of riots, which became more frequent and more serious as the use of machinery spread. In Lancashire, where machinery had developed most quickly, these disturbances in 1779 became really alarming. Wedgewood, who was in the district when the riots broke out, wrote in one of his letters an account of them which has the value of evidence given by an eye-witness: In our way to this place [Bolton], a little on this side [of] Chowbent, we met several hundred people in the road. I believe there might be about five hundred; and upon inquiring of one of them the occasion of their being together in so great a number, he told me they had been destroying some engines, and meant to serve them all so through the country. . . .

Two of the mob were shot dead upon the spot, one drowned and several wounded. Accordingly, they spent all Sunday and Monday morning in collecting firearms and ammunitions and melting their pewter dishes into bullets. They were now joined by the Duke of Bridgewater's colliers and others, to the number, we were told, of eight thousand, and marched by beat of drum and with colours flying to the mill where they had met with a repulse on Saturday. . . . The mob completely destroyed a set of mills valued at £ 10,000. This was Monday's employment. On Tuesday morning we heard their drums at about two miles' distance from Bolton, a little before we left the place, and their professed design was to take Bolton, Manchester and Stockport on their way to Cromford, and to destroy all the engines not only in these places, but throughout all England.[3]

These violent riots were the last resort of the industrial workers. They tried to have Parliament pass laws prohibiting the use of cotton-spinning machinery and later, wool-combing machinery, arguing that the loss of jobs would turn hitherto self-sufficient workers into public charges. Machinery was looked upon as depriving the worker of his only capital—his labor.

It appears to the petitioners that one machine only, with the assistance of one person and four or five children, will perform as much labour as thirty men in the custom-

3. Paul Mantoux, *The Industrial Revolution in the Eighteenth Century*. (New York: Macmillan 1961), 401–2.

8

ary manual manner. . . . And in consequence of the introduction of the said machine the great body of woolen
manufacturers will almost immediately be deprived of
their business and employ, the whole trade will be en-
grossed by a few powerful and wealthy adventurers, and
after short competition the surplus profit arising from the
annihilation of manual labour will be transferred into the
pockets of foreign consumers.[4]

A secondary cause of opposition to the machinery was the workers' hatred of
the factory. The Industrial Revolution called for a new kind of discipline which
the workers were not used to:

Children, women, and men were subjected in factories to
conditions and disciplines not known to them before. The
buildings were often of hasty or flimsy construction, assuring many accidents and much disease. . . . Employers
argued that the proper care of the machinery, the necessity of co-ordinating different operations, and the lax habits of a population not accustomed to regularity or speed
required a rigorous discipline if confusion and waste were
not to cancel profits and to price the product out of the
market at home or abroad. . . .
Under the guild system the workers were protected by
guild or municipal ordinances, but in the new industrial-
ism they had little protection by the law, or none at all.
. . . [T]he employers convinced Parliament that they
could not continue their operations, or meet foreign competition, unless wages were governed by the laws of supply and demand.[5]

To the worker, proud of his creation by hand, his displacement by the machine was traumatic. Recognizing full well the leisurely and self-set pace of the
workers' manual production, the employers, who were investing sizeable amounts
of capital, insisted that the workers assemble under one roof, come to work on
time, and stay on the job for the working day. What gave the Industrial Revolution a bad name, however, were not the necessary demands of factory production,
but the *excesses* committed in their name by the employer-investor:

Here we come to the real cause of the evils attributed to
machine industry, namely the absolute and uncontrolled

4. *Journals of the House of Commons*, XLIX, 545–46.
5. Will and Ariel Durant, *op. cit.*, 678–79.

power of the capitalist. In this, the heroic age of great un-
dertakings, it was acknowledged, admitted and even pro-
claimed with brutal candour. It was the employer's own
business, he did as he chose and did not consider that any
other justification of his conduct was necessary. He owed
his employees wages, and once those were paid the men
had no further claim on him: put shortly, this was the attitude of the employer as to his rights and his duties.[6]

Despite the creative role of the capitalist through the ages in financing the
mechanization of industry and agriculture, rationalizing distribution, and producing "such a flow of goods from producer to consumer as history has never seen
before," the Durants find that "they do not explain why history so resounds with
protests and revolts against the abuse of industrial mastery, price manipulation,
business chicanery, and irresponsible wealth."[7] From the earliest days, not just of
the Industrial Revolution, but of human history itself, people have tried socialistic
experiments in which the state, not the capitalist, organizes, owns, and operates
the means of production: Sumeria (2100 B.C.), Babylonia (1750 B.C.), Egypt (323
B.C.–30 B.C.), Rome (A.D. 301), China (A.D. 9–23; 1068–85), Peru (1250),
Uruguay (1620–1750), Germany (1530–1622), and Russia (1917).[8]

What the Durants call the "abuse of industrial mastery" was most vivid during the early days of the Industrial Revolution, and they attribute to the brutality
of these early days the later appeal of Marxism:

The socialist agitation subsided during the Restoration,
but it rose again when the Industrial Revolution revealed
the greed and brutality of early capitalism—child labor,
women labor, long hours, low wages, and disease—
breeding factories and slums. Karl Marx and Friedrich
Engels gave the movement its Magna Carta in the *Com-
munist Manifesto* of 1847, and its Bible in *Das Kapital*
(1867–95). They expected that socialism would be effected first in England, because industry was there most
developed and had reached a stage of centralized management that seemed to invite appropriation by the government. They did not live long enough to be surprised by
the outbreak of Communism in Russia.[9]

6. Paul Mantoux, *op. cit.*, 417.
7. Will and Ariel Durant, *The Lessons of History* (New York: Simon and Schuster, 1968), 59.
8. *Ibid.*, chapter IX.
9. *Ibid.*, 65.

The *Manifesto* made its appearance in German two years earlier, in 1845. That would date the *Manifesto*'s clarion call for the workers of the world to rise up in revolt seventy-six years after the act of Parliament decreeing death for riots against machinery. The battle cry of the English workers who repeatedly rioted in spite of the threat of death in the second half of the eighteenth century could well have given rise to the subsequent rallying cry of the *Manifesto*: "Workers of the world unite; you have nothing to lose but your chains!"[10] To the English workers, though, their world was their particular industry. The chains they would lose would be their jobs.

0.0 "Cotton Manufacturers of the Kingdom . . . Carried to a Height Both of Excellence and Extent"[11]

Workers who repeatedly rioted to fight for their jobs in the cotton industry in the early days of the Industrial Revolution found by the 1790s that their worst fears were proven unfounded. They found employment "in the progress," and cotton manufacture excelled in quality and quantity.

0.0 Increased Productivity

Earlier we saw the workers in woolen manufacturing petition Parliament in 1794 stating that one machine would do the work of thirty workers and that consequently the displaced workers would be ruined. They said: "[I]t is with the most heartfelt sorrow and anguish the petitioners anticipate that fast approaching period of consummate wretchedness and poverty, when fifty thousand of the petitioners, together with their distressed families, by a lucrative monopoly of the means of earning their bread, will be inevitably compelled to seek relief from their several parishes."[12] This petition met with a favorable response initially, but then the employers put forth their story.

The employers argued that Cartwright's invention of wool-combing enriched the public. "The public has become possessed of a very valuable improvement in the combing of wool by machinery."[13] They asserted that the machinery would reduce the price of wool by 75–80 percent and would save more than one million British pounds a year. They argued, in a word, *productivity*.

> If in the progress of time the machine combing, being
> unrestricted by law should supersede the use of hand
> combing, the saving to the national produce would, on a

10. Friedrich Engels, *Die Lage der arbeitenden Klasse in England* (1845).
11. *Journals of the House of Commons*, XLIX, 546.
12. *Ibid.*, 545.
13. *Ibid.*

11

moderate calculation, greatly exceed £1,000,000 per annum; and, if the machine combing is suppressed, the national manufacturers must in consequence be charged
with this additional load.[14]

Not only would the machinery enormously improve productivity, then, but
failure to use the machinery would levy severe penalties on the nation's industry.
"If the petitioners should be compelled to abandon the combing of wool by machinery, they will severely be subjected to the ruinous obligation of producing their
yarns at an expense of from £1,500 to £2,000 per annum more than the production of the same yarn will cost by machinery."[15]

0.0 Increased Employment

Perhaps the clincher in the employers' argument was that the much-feared unemployment had not materialized in the cotton industry, where a much larger number
than that in the woolen industry was at risk from the machinery.

> The good policy of the law in leaving manufactures to
> their natural course was strikingly evinced in the business
> of cotton, in the spinning of which the application of ma-
> chinery threatened to deprive of their maintenance a much
> more considerable body of artizans, while in the progress
> the artizans found employment and the cotton manufac-
> tures of the Kingdom were carried to a height both of
> excellence and extent, and the national wealth was multi-
> plied, beyond all example. The woolen manufacturers
> may, in all probability, from the same causes, reach the
> same height of prosperity and excellence, if unchecked
> by prohibition laws.[16]

It should be noted that the employers argued that the cotton workers found employment "in the progress" of mechanization. In other words, it took some time
before the machinery stimulated enough employment to offset the initial lay-off of
workers.

The inventor of the "spinning jenny," James Hargreaves of Blackburn, Lancashire, had to flee for his life to Nottingham when the workers broke up his machines. Although he invented the jenny in 1765, he did not make machines for
sale until 1767, "and at once fell victim to that unpopularity which inventors in
those days seldom escaped. Blackburn workers broke in his door and smashed his

14. *Ibid.*, 546.
15. *Ibid.*
16. *Ibid.*

machines."[17] In Nottingham, where he flew for safety, there was a shortage of workers, which allowed him to sell his spinning jennies. In 1788, ten years after his death, "it was reckoned that there were no fewer than twenty thousand of these machines in England, of which the smallest could do the work of six or eight spinners."[18] Even Lancashire, which originally drove him out, took up the jenny. "In Lancashire they spread with astonishing rapidity, and in a few years had completely ousted the spinning wheel."[19]

"In the progress" of adopting the cotton machinery, then spanned but a decade. Well could the woolen-industry employers point with pride to what was accomplished in cotton and thereby persuade Parliament that, in time, woolen manufacturing could reach a similar prosperity.

0.0 "East-India Trade . . . a Very Likely Way of Forcing Men Upon the Invention of Arts and Engines"[20]

We saw that the "demands of expanding markets at home and abroad" required far greater capability of production of goods and services than what the guilds were able to provide. One pamphleteer, with great foresight, described the role of foreign trade in the Industrial Revolution more than three-quarters of a century before the event:

> That this thing may not seem a Paradox, the EAST-INDIA Trade may be the cause of doing things with less Labour, and then tho' wages shou'd not, the price of Manufactures might be abated. . . . In like manner of any English Manufacture perform'd by so many Hands, and in so long a time, the price is proportionable, if by the invention of an Engine, or by greater order and regularity of the Work, the same shall be done by two-thirds of that number of Hands, or in two-thirds of that time; the labour will be less, the price of it will be also less, tho' the Wages of Men shou'd be as high as ever. . . . And thus the EAST-INDIA Trade by procuring things with less, and consequently cheaper labour, is a very likely way of forcing Men upon the invention of Arts and Engines, by which other things may also be done with less and

17. W. A. Abram, *Parish of Blackburn, County of Lancaster. A History of Blackburn. Town and Parish*, Blackburn, 1877.

18. *An Important Crisis, in the Callico and Muslin Manufactory in Great Britain, explained*, London, 1788.

19. John Kennedy, "A Brief Memoir of Samuel Crompton," *Memoirs of the Literary and Philosophical Society of Manchester*, Second Series, Vol. 330, Manchester, 1819.

20. *Considerations upon the East-India Trade*, attributed to Sir Dudley North, London, 1701.

> cheaper labour, and therefore may abate the price of
> Manufactures.[21]

The author of the preceding pamphlet, believed to be Sir Dudley North, was responding to a serious controversy of the day, namely, import of goods from India into England. What North set out to do was to show that by buying goods which could be bought more cheaply from abroad, English labor could be saved to create new industries, or better distribution of functions within existing industries.

Three-quarters of a century later, Adam Smith would express his famous theory that "[t]he division of labour is limited by the extent of the market."[22] That theory included in its sweep both domestic and foreign markets, including faraway India.

Foreign trade has been of paramount importance to the island-nation that has had to import virtually all of its necessities from abroad. The expansion of her territory revolutionized her trade.

> Command of the seas facilitated the conquest of colonies.
> Canada and the richest parts of India fell to England as
> fruit of the Seven Years' War. Voyages like those of
> Captain Cook (1768–76) secured for the British empire
> islands strategically useful in war and trade. Rodney's
> victory over de Grasse (1782) confirmed British dominion
> over Jamaica, Barbados, and the Bahamas. New Zealand
> was acquired in 1787, Australia in 1788. Colonial and
> other overseas trade gave British industry a foreign mar-
> ket unrivaled in the eighteenth century.[23]

If Dudley North's focus was on saving English labor, which could then be devoted to inventing "Arts and Engines," that is, machinery, the expansion of English territory focused attention upon the expanding markets. "The English economy was not only spared the ravages of soldiery, it was nourished and stimulated by the needs of British and allied armies on the Continent; hence the special expansion of the textile and metallurgical industries, and the call for machines to accelerate, and for factories to multiply, production."[24]

British trade was a key factor in judicial decisions pertaining to riots against the machinery in the early days of the Industrial Revolution.

> Resolved that the sole cause of great riots was the new
> machines employed in the cotton manufacture; that the

21. *Ibid.*, 65–67.
22. Adam Smith, An Inquiry into the Nature and Causes of the Wealth of Nations, (ed. Edwin Cannan), Modern Library, New York, 1937 p. 17.
23. Will and Ariel Durant, *Rousseau and Revolution, op. cit.*, 669.
24. *Ibid.*

country notwithstanding has greatly benefitted by their erection; that destroying them in this country would only be the means of transferring them to another country, and that, if a total stop were put by the legislature to their erection in Britain, it would only tend to their establishment in foreign countries, to the detriment of the trade of Britain.[25]

This resolution, or judgment, was given on 11 November 1779. Whether or not the Justice of the Peace had read Adam Smith's *Wealth of Nations*, published three years earlier, he showed an awareness of the role of machinery in foreign trade. In a sense, he went beyond dispensing a judicial ruling. He suggested to the Parliament that they would do something to the detriment of British trade if they prohibited cotton machinery. The Parliament did not, and in another nine years, twenty thousand spinning jennies were whirring away in England.

0.0 From Resistance to Rapport

This brief survey of the early days of the Industrial Revolution offers some clues as to what types of response may be expected in the early days of the Robotics Revolution.

0.0 Capital Concerns

It is easy to recognize the fierce resistance of the workers to the machinery during the birth of the Industrial Revolution. There was a tremendous but less obvious readjustment in the role of capital as well.

We observed that inventions by themselves do not create innovations, that the application of inventions to a social process, such as manufacture of goods and services is not automatic. The British historian, Arnold Toynbee, points out that inventions in ancient Rome fell by the wayside because the political powers were either indifferent or hostile to them:

> Certainly the Roman imperial government did not even realize, at any stage of its history, that technology, as exemplified in *Hero of Alexander*'s *Invention of a Turbine Engine*, could have solved the Hellenic universal state's intertwined problems of finance and defence. And in the western provinces in the fourth century of the Christian Era, when the Empire was fighting for survival there, no attention was paid to possibilities of dealing

25. Sidney and Beatrice Webb, "Webb Trade Union Collection" (London: London School of Economics) Collection E, Section A, Textiles I, 277.

with manpower shortage and with defence logistics by mechanization, though a set of projects for this was published in an anonymous memorandum *De Rebus Bellcis.*[26]

Whether or not the turbine engine would have solved the problems of the day is beside the point. The fact is that the invention had no sponsor. It did not have "projectors" who would convert it into a social innovation. Unless entrepreneurs of the Robotics Revolution undertake to acquire capital for both the present and/or potential robotics inventions, the inventions may well languish.

0.0 Market Concerns

Even if the capital were to be found for Robotics Revolution inventions, the success of an invention depends on the market for the products and services it generates when applied to meet the needs and wants of society. The entrepreneurs of the Industrial Revolution were able not only to meet, but more importantly, to "anticipate or create demand." Since robotics inventions are either futuristic or barely emerging, the task of anticipating successfully the demand of a public for the unborn product of an unknown invention is more than a trifle complicated. Going one step further, the robotics entrepreneur actually has to create such a demand, a task requiring Herculean imagination and unerring instincts about the volatile tastes and temperament of the consumer.

0.0 International Dimensions

In the case of the Industrial Revolution, "a foreign market unrivaled in the eighteenth century" in large measure accounted for success. The market preceded the mechanization; the territory was assured beforehand for the technology. The entrepreneur of the Industrial Revolution could without hesitation proceed to "organize machines and men to produce for markets in every quarter of the globe."

Today, such an assured territory has not paved the way for the robotics entrepreneur. He has to create a territory, inch by inch, at home and abroad. The imperatives of international markets are well recognized by Japan, where the automakers are vigorously pursuing international links.

> "What you are seeing now is a crucial period when sales and production strategy undergoes a drastic change," says a Nissan [which has just under 30 percent of the domestic market, second to Toyota with 40 percent] spokesman. "The domestic market is going to get more difficult from now on. And there is a growing acceptance that the future

26. Arnold J. Toynbee, *A Study of History,* a one-volume condensation of a 10-volume work of the same title published in 1935 (Oxford, 1972), 63.

lies in more overseas production facilities and tieups with other makers for survival. We no longer believe we can go it alone in the world. . . ."

Nissan, for example, is building a truck assembly plant in Smyrna, Tennessee. It has its own vehicle assembly operations in Mexico and Australia and is about to announce a decision on a passenger car assembly plant in Britain. . . .

Toyota and Nissan, for example, are competing to be chosen as partner with a Taiwanese company to develop the island's nascent auto industry. Mitsubishi Motors has signed up to help the expansion-minded South Koreans, while Suzuki motors is to cooperate with India in small car production.[27]

0.0 Unemployment Fears

We saw that the workers, fearing for their jobs, smashed the Industrial Revolution inventions, pleaded with Parliament to prohibit the machinery, rioted as mobs numbering five hundred to eight thousand, demolished in one day mills valued at £10,000, and vowed to destroy "all the engines . . . throughout all England."

Undoubtedly, unemployment will be a significant, if not a paramount, concern in the Robotics Revolution. In Japan, the problem will be different from that in the United States because a fourth or more of Japanese employees work for the same employer all their life. Because of the life-long commitment to each other, the employees feel less threatened by labor-displacing devices, trusting that their employers will keep them gainfully occupied. Nevertheless, Professor Hasegawa finds unemployment a problem in the Robotics Revolution:

> When a number of robots are introduced [sic], an unemployment problem will break out among the unskilled workers who cannot do any other sophisticated operations. . . .
>
> Even the skilled workers who have been working with pride as specialists will be treated as equal as the unskilled workers if he would not try to learn how to operate robots or get another job and try to be a new skilled worker again. . . .
>
> As the robotization proceeds, un-employment problems will break out, while engineers or skilled workers neces-

27. Geoffrey Murray, "Will GM-Toyota Link Shake Up the Auto World?" *The Christian Science Monitor*, 13 May 1982, 80.

sary for operating robots and taking care of them will become insufficient. . . .

Up to this time, there was a flow from the countries where the labor cost was high to the countries where the labor cost was low, which is so-called catch-ups of developing countries. However, as the robotization proceeds, in developed countries many types of goods will be made automatically which are of high quality and low cost. So the flow will stop and there will be a possibility that the income difference between developed countries and developing countries will become bigger than before.[28]

Professor Hasegawa highlights the unemployment consequences of the Robotics Revolution for the skilled workers as well as for the unskilled. The Robotics Revolution will demand workers with different skills, forcing the skilled worker to be relocated, retrained, or unemployed. Unemployment, which signifies an excess of workers over jobs, will coexist, then, with shortages of workers trained in robotics skills.

What about the international impact? Professor Hasegawa foresees robotics slowing the current flow of products from developing to developed countries. While the initial cost of invesment is high for robotics, the unit cost of production is lower, making it uneconomical to depend on cheaper labor in developing countries. As a consequence, the earnings of developing countries from their labor will decline, and the difference in income between the developed and developing countries will increase.

The U.S. Labor Department estimates that robotics could displace seven million jobs in the next decade or two. Currently, the number employed in manufacturing is twenty million. That is a whopping 35 percent unemployment. The December 1982 recession, with 10.8 percent unemployment—a post-Depression high—was less than a third of that rate.

Softening the spectre of such a massive shrinkage in current manufacturing jobs is the necessary spreading out of the introduction of robotics over time and space. It is quite unlikely that all industries will embrace robotics to the same extent, at the same time, and all over the country. Further, the robotics capabilities that are required in different manufactures are not likely to become commercially viable all at once.

On the other hand, if domestic as well as international markets shrink with the advance of robotics elsewhere, which happened when Japan carved out a fifth of the U.S. domestic market for automobiles, there would be renewed pressures on the U.S. industry to embrace robotics in self-defense. There would be a double

28. Yukio Hasegawa, "How Robotics Have Been Introduced into the Japanese Society," International Symposium on Robots and Man (Osaka, Japan: 19 August 1982), 9.

crunch—from the lost domestic market and from the lost domestic jobs in manufacturing.

0.0 Dealing with Fears

The Robotics Revolution brings with it fears of the unknown not dissimilar to those caused by the Industrial Revolution. With the benefit of hindsight, is it possible to identify the elements that enabled the industrial workers of eighteenth century England to drop their resistance and eventually embrace the very machines that had so threatened their livelihood?

Around the time of the 1779 riots, one of the magistrates, Dorning Rasbotham, wrote a pamphlet under the nom de plume "A Friend of the Poor." He pointed out that the troubles of the day were transistory and that continued resistance would mean economic death.

> All improvements in trade by machines do at first produce some difficulties to some particular persons. . . . About ten years ago, when the spinning jennies came up, old persons, children and those who could not easily learn to use their own machines did suffer for a while. . . . Was not the first effect of the printing press to deprive many copyists of their occupation? What mean those riots and tumults which we saw a few months ago? What mean the petitions to Parliament to suppress or tax the machines? We might just as well ask to have our hands lopped off or our throats cut.[29]

The key is the recognition that "some difficulties to some particular persons" is an inevitable price to pay for the new era. While that point was obvious to the magistrate who had to sentence the rioters for the destruction of property, it was obviously not acceptable to the workers who found in the machinery the depriver of their livelihood. A decade is but 5 percent of the two-hundred-year Industrial Revolution, but 40 percent of the twenty-five-year working life of a breadwinner. Once the job was lost, there was no other trade the worker could learn. No wonder, the Luddites destroyed the machines that would condemn them to economic death.

In addition to the judges, the lawmakers of the day accepted the inevitability of hardships as a consequence of the introduction of the machinery. Another magistrate, rendering judgment on rioters, told the legislature that it should not accede to the petition by the workers to ban cotton machinery because to do so would be

29. (Dorning Rasbotham) *Thoughts on the Use of Machines in the Cotton Manufacture, addressed to the Working People in that Manufacture and to the Poor in General, by a Friend of the Poor*, Manchester, 1780, 9, 11, 20.

detrimental to foreign trade. For a country depending on foreign trade for a significant segment of her national income, this would be tantamount to economic suicide.

What made the "difficulties to some particular persons" acceptable to the nation's leaders, then, was the paramountcy of foreign trade. Put negatively, if machinery were banned in England by Parliament, "it would only tend to their establishment in foreign countries, to the detriment of the trade of Britain." Put positively, the unrivaled eighteenth century foreign market produced a steady and steadily increasing demand for the products of the machinery. The territory, the market, made the *technology*, the machinery, not only possible but imperative. Produce or perish!

The market was necessary, but not sufficient, to overcome the opposition of the worker. Use of machinery was extremely risky—first, the high cost of purchase; second, the physical destruction by the workers; and third, the sabotage of the production by the workers—the word itself being derived from the French word for shoes, which were thrown by the workers into the machines to disrupt and destroy. Yet, in spite of all of this, there were entrepreneurs, "projectors," who would not only meet demand, but also create it.

The major elements that changed the workers' attitude, then, were, the market which generated enormous demand for the products, the real threat of national ruin if improved productivity through mechanization was not adopted, the acceptance by the judges and the lawmakers of the inevitability of hardships, and the increased earnings emerging as a clear possibility.

England's colonial conquests did bring her an enormous market for her products. But the *sustained* success of the Industrial Revolution in maintaining and expanding both domestic and foreign markets came from the productivity of the Industrial Revolution, the output per worker and/or output per hour. England, being the first country to embrace the machines of the Industrial Revolution, gained a headstart over other countries in productivity, and she could keep her lead by improving her processes and products. So too with the Robotics Revolution, the single determining factor is likely to be productivity. As the revolution in robotics spreads, the issue will be productivity, both at home and abroad.

2

The Sources and Signals of Declining U.S. Productivity

OVERVIEW: The Bureau of Labor Statistics defines productivity as labor productivity—the value of goods and services produced per hour of labor. It is clear that, by any account, the rate of increase in labor productivity in the U.S. has been decreasing.

Among the primary sources of productivity decline are: *capital* (20%), *sectoral shifts* (14%), *research* (11%), *regulation* (8%), *labor* (7%). Altogether, these account for 60 percent. These percentages are unweighted averages of several expert estimates that sometimes vary widely.

Other specific factors chosen by different authors account for 30 percent, the remaining 10 percent being residual.

While the 2.1 percent increase in labor force in the 1972–79 period was 525 percent of the 0.4 percent increase in the 1948–65 period, capital investment grew only 6 percent. It represented 9.5 percent of the gross national product in the 1948–65 period and 10.3 percent in the period 1973–79. Why? Economic regulation, under the rubric of which should be included all rules and regulations that stimulate or stifle savings by individuals and corporations, deposits of these savings in banks and other financial institutions, the lending of these savings by banks to business, and the investment by business in physical stock, which multiplies human production efforts (capital outlay), and in intellectual stock, which multiplies human production effort (research and development).

The contrast of declining American productivity with rising Japanese productivity is dramatic. Associating the productivity performances with the respective national flags, we can talk of the Falling Star and the Rising Sun. "Growth agreed upon and growth pursued" characterizes Japanese production and enhances productivity, the elements of which definitely merit further scrutiny.

0.0 The Decline and Fall of U.S. Productivity

The near-universal concern with declining labor productivity in the U.S. is dramatically underscored when a Japanese economist opens his paper, "Prospects for the 1980s—a Japanese View," with a direct contradiction of this conventional wisdom:

> The United States is not a declining economic power plagued by a secular fall in productivity. . . . While productivity in the U.S. economy as a whole at the beginning of 1979 was little changed from its 1973 level, productivity in the manufacturing sector increased by over 10 percent. . . . And it directly challenges the currently popular belief that America is a declining world economic force and supports the view that in the 1980s American goods will become increasingly more competitive in international markets.[1]

Is U.S. productivity declining? Is Mr. Nakame just being polite? Is the increase in manufacturing productivity enough to make American goods competitive in the 1980s?

0.0 Wanted: Concepts Beyond Labor Productivity

It is critical to define our terms of reference when we speak of productivity. Almost always, we mean labor productivity, to the exclusion of productivity of capital and technology. Capital and technology, though, are particularly significant in the area of robotics and artificial intelligence.

Most often productivity is figured as a ratio of output (GNP) to input (number employed or number of paid hours worked). It is quite conceivable that this output/input ratio is too crude a measure to reflect the significant role of non-labor elements in producing a larger output from the same input. In particular, robot factories can send the labor productivity figure sky-high if the three or four people required to check on the robots alone are counted in the denominator. What we are really measuring is technology productivity rather than labor productivity.

It should be noted that at least three major factors are camouflaged by our ascribing the technology productivity of robotization to labor: substantial capital investment requirements, significant initial unemployment, and indispensable "projectors," the risk-taking entrepreneurs.

---

1. Tadashi Nakamae, "Prospects for the 1980s—A Japanese View," in *The International Politics of Surplus Capacity,* ed. Susan Strange et al. (London: George Allen & Unwin, 1981), 169.

The Sources and Signals of Declining U.S. Productivity

0.0 The Triad of Japanese Productivity

We may consider the revolution in robotics in the early 1980s as somewhat parallel to the computer revolution of the 1950s. In thirty years, the computer has gone from the first generation to very nearly the fifth. On a similar scale of development, robotics could reach the fourth or fifth generation by the turn of the century.

The Japanese are committed to capturing the coming revolution in robotics. Indeed, the much-heralded Fifth-Generation computers that the Japanese industry-university-government consortium is working on is representative of the concentration of resources in what Japan considers a number one priority.

The battle cry of the revolution in robotics is, "Productivity." While the real gross product per labor hour in the U.S. dropped from an average of 3.5 percent per year in the 1948–66 period to −0.9 in 1980,[2] it rose in Japan at an annual average rate of 7.3 percent.[3] Small wonder the composite yield in high-technology electronics in the U.S. is 12 percent, while in Japan, at 85 percent, it is more than seven times as high.[4]

And this is just the *start* of the robotics race.

Number 2 must try harder. It must try harder to set and reach goals. The U.S. does not have to copy Japan in productivity; it can set its own goals. But the United States must know what it is competing against and what resources it has available to throw into the competition. We need to take a hard look at what factors have caused the decline in U.S. productivity and learn what factors have caused the growth of Japanese productivity. We will discuss the latter in chapter 3, but first the productivity definitions and a look at the U.S.

0.0 Labor Productivity Definitions—the U.S. and Japan

The Bureau of Labor Statistics defines productivity as labor productivity—the value of goods and services produced per hour of labor. Output is "constant dollar domestic product" produced in a given period. "Compensation per hour" includes wages and salaries of employees plus employers' contributions for social insurance and private benefit plans. The data also include an estimate of wages, salaries, and supplementary payments for the self-employed. "Real compensation per hour" is compensation per hour adjusted by the Consumer Price Index for All Urban Consumers.[5]

The Japanese have a similar definition of productivity. In a study of factors which help or hinder improvement of productivity, Professor Ryokichi Hirono

---

2. John W. Kendrick, "Survey of Factors Contributing to the Decline in U.S. Productivity Growth", in *The Decline in Productivity Growth* (Boston: the Federal Reserve Bank of Boston, June 1980), 3 (hereafter cited as *The Decline*).

3. O. Friedrich, "The Robot Revolution," *Time*, Vol. 116, No. 23, 1980, 72–83.

4. Jay C. Lowndes, "Productivity Growth is Key Goal," *Aviation Week and Space Technology*, 2 August 1982, 42.

5. U.S. Department of Labor, *Monthly Labor Review*, January 1983, 99.

23

talks about two types of labor productivity: output per worker and value added per worker. He also recognizes the role of capital and equipment in productivity:

> Presumably, there are three ways of increasing the output or value added per worker: increasing the physical productivity per worker, the value of productivity per worker with or without increasing physical productivity . . . [and] in good times, when employment and output both were on a rapid expansion, the rise in the physical productivity per worker would result mainly from increases in capital investment and capacity utilization.[6]

The Productivity Research Institute of the Japan Productivity Center, established in 1955, considers physical productivity, measured in units of resources, as indicative of the degree of *technological rationality* of management activity, while the value-added productivity, measured in net output value-added by manufacture, as indicative of the degree of *economic rationality* of management activity.[7]

0.0 U.S. Labor Productivity Decline

We see by studying the annual indexes of U.S. productivity, provided by the Department of Labor, that the rate of growth in productivity is smaller in recent years. In 1960, manufacturing productivity (labor productivity in manufacturing) rose to 60.0 from 49.4 in 1950, or a 20 percent increase. In 1970, it rose to 79.1 from 60.0 in 1960, or a 33 percent increase. In 1980, it rose to 101.7 from 79.1 in 1970, or a 29 percent increase. During the 1950–80 period, the average annual increase in manufacturing productivity was 3.5 percent per year. From 1980 to 1981, however, the increase was only 2.7 percent. (See table 2.1.)

How does the 3.5 percent figure compare with the corresponding figures in nonfinancial corporations and the nonfarm business sector?

Both were smaller. The nonfinancial corporations registered an average annual increase of 2.3 percent during the 1960–80 period (prior figures not being available), while the nonfarm business sector registered 2.5 percent.

At the end of the thirty-year period, manufacturing productivity rose by 2.7 percent in 1981 over 1980, the productivity of nonfinancial corporations rose 1.6 percent, and nonfarm business sector productivity 1.4 percent.

It is clear that by any account, the increase in labor productivity in the U.S. has been decreasing. This is reflected in the title of a conference held by the Federal Reserve Bank of Boston: "The Decline in Productivity Growth."

---

6. Ryokichi Hirono, *Factors Which Hinder or Help Productivity Improvement: Country Report Japan* (Tokyo: Asia Productivity Organization, 1980), 55–56.
7. Kohey Goshi, "Analysis of the Value Added" (Tokyo: Japan Productivity Center, January 1980), 1.

Table 2.1
Annual Indexes of Productivity
Output per Hour of All Persons

	1950	1960	1970	1975	1976	1977	1978	1979	1980	1981
Manufacturing	49.4	60.0	79.1	93.4	97.5	100.0	100.9	101.5	101.7	104.5
Nonfinancial corporations	n.a.	68.0*	87.4*	95.5*	98.2*	100.0	100.9*	100.7*	100.3*	102.0*
Nonfarm business sector	56.3	68.3	86.8	94.7	97.8	100.0	100.6	99.3	98.5	99.9

Source: Monthly Labor Review, January 1983, 99.
*Revised

One of the speakers who presented papers, John W. Kendrick, uses *real product per unit of labor* in the U.S. Domestic Business Economy to measure productivity. He uses different periods from those of the Department of Labor cited in table 2.1, but the conclusion is the same. During the 1948–66 period, real product per unit of labor registered an average annual percentage rate of change of 3.5 percent; during 1966–73, 2.1 percent, and during 1973–78, 1.1 percent.[8]

0.0 Sources of Productivity Decline

Kendrick accounts for the 2.4 percentage point decline (from 3.5 percent in the first period, 1948–66, to 1.1 percent in the third, 1973–78) in terms of reduced contribution from five sources (table 2.2). These sources are presented in table 2.3, reproduced from Nordhaus and supplemented by the author with Nordhaus's own set of sources. Nordhaus measures productivity decline in a manner quite similar to that of Kendrick: "The 'slowdown' is the difference in the growth rate of productivity per hour worked from the period 1948–65 [Kendrick uses 1948 –66] to the period 1973–79 [Kendrick uses 1973–78]. Output is gross product originating in the private business sector."[9]

What are the primary sources of decline in productivity?

We compute each source's contribution to the total productivity decline based on the entries in table 2.3:

1. **Capital** accounts for 16% (Kendrick), 35% (Norsworthy, Harper, Kunze), 7% (Denison), 29% (Clark) and 12% (Nordhaus) of the respective total productivity decline. The unweighted average is 20%.

Table 2.2
Sources of Productivity Decline

	Annual Average Percentage Rates of Change			
	1948–66	**1966–73**	**1973–78**	**1973–78 – 1948–66**
Real product per unit of labor	3.5	2.1	1.1	− 2.4
Capital/labor substitution	0.7	0.5	0.3	− 0.4
Advances in technological knowledge	1.4	1.1	0.8	− 0.6
Resource reallocations	-	− 0.1	− 0.2	− 0.2
Volume changes	0.4	0.3	0.2	− 0.2
Net government impact	-	− 0.1	− 0.3	− 0.3

Source: John W. Kendrick, "Survey of Factors Contributing to the Decline in U.S. Productivity Growth," in *The Decline in Productivity Growth* (Boston: the Federal Reserve Bank of Boston, June 1980), 3.

8. Kendrick, *op. cit.*, 3.
9. William D. Nordhaus, "Policy Responses to the Productivity Slowdown," in *The Decline*, op. cit., 153.

Table 2.3

Changes in the Rate of Growth of Labor Productivity. Pre-1965 to post-1972

	J. Kendrick	Norsworthy, Harper, Kunze		E. Denison	Z. Griliches	P. Clark	P. Clark	L. Thurow	W. Nordhaus	Miscellaneous
Sector	Private Business	Private Business	Private Nonfarm Business	Total Economy	Manufacturing	Private Nonfarm Business	Private Non-farm, Non-residential Business	Private Nonfarm Business	Private Business	Private Business
Output measure	Gross Dom. Inc.	Gross Dom. Inc.	Gross Dom. Inc.	Net Nat'l. Inc.	Gross Output	Gross Dom. Inc.	Gross Dom. Inc.	Gross Dom. Inc.	Gross Output	
Periods studied	1948–66 & 1972–78	1948–65 & 1973–78	1948–65 & 1973–78	1953–64 & 1972–76	1959–68 & 1969–77	1948–65 & 1973–76	1848–65 & 1973–78	1948–72 & 1972–78	1948–65 & 1973–79	
Total decline	−2.40	−2.12	−1.68	−2.64			−1.83	−0.40	−2.50	
Cyclical	−0.30			−0.05					−0.30	
Trend						−1.67			−2.20	
Capital	−0.40	−0.74†	−0.57†	−0.17		−0.4 to −0.97‡	−0.54†		−0.30	
Labor	+0.10	−0.28	−0.18	−0.14					−0.10	
Energy		−0.18‡	−0.18‡	−0.10	−0.10 to −0.40	+0.04			−0.20	−0.6 (Jorgenson-Hudson) −1.3 (Rasche-Tatom) −0.2 (G. Perry)
Regulation	−0.30	−0.09	−0.08	−0.27				−0.20*	−0.20	
Research	−0.60			−0.10					−0.10	
Sectoral shifts	−0.50			−0.27				−0.60**	−0.30	−0.10 (CEA)
Other factors	−0.30	−0.83	−0.67	−1.54		−0.67 to −1.28	−1.29		−1.00	

†Includes utilization effect. ‡Manufacturing. *Mining only. **Output composition change.
Source: William D. Nordhaus, "Policy Responses to the Productivity Slowdown," in *The Decline in Productivity Growth*, (Boston. the Federal Reserve Bank of Boston, June 1980), 152–53.

2. **Sectoral Shifts** account for 21% (Kendrick), 10% (Denison), and 12% (Nordhaus): average 14%.
3. **Research** accounts for 25% (Kendrick), 4% (Denison), and 4% (Nordhaus): average 11%.
4. **Regulation** accounts for 12% (Kendrick), 4% (Norsworthy, Harper, Kunze), 10% (Denison), 8% (Nordhaus): average 8%.
5. **Labor** accounts for 13% (Norsworthy, Harper, Kunze), 5% (Denison), 4% (Nordhaus): average 7%.

Thus, we have the unweighted average contributions to productivity decline due to capital (20%), sectoral shifts (14%), research (11%), regulation (8%), and labor (7%), which together account for 60%. Other specific factors chosen by different analysts account for 30%, the remaining 10% being residual.

Three primary sources of decline in productivity, then, are capital, sectoral shifts, and research. Together, they account for nearly half the productivity decline. Directly affecting capital formation and investment in plant and equipment, as well as research and development, is "economic" regulation, which determines the attractiveness of investment in business equipment (capital) and knowledge (research).

0.0 Decline Source: Capital Accumulation
Richard W. Kopcke concludes that much of the 1965–78 slump in productivity and potential GNP growth resulted from a slower rate of capital accumulation:

> Whereas the average annual growth of the stock of equipment exceeded [labor] hours by 2.8 percentage points before 1965, the growth of equipment, on average, surpassed hours by only 2.2 percentage points thereafter. From 1973 to 1978 the annual expansion of equipment exceeded hours by only 1 percentage point. . . .
>
> The cost of capital services for producers' durable equipment for non-farm, non-residential business fell 7.3 percent per year from 1950 to 1965; since then, capital costs fell only 5.7 percent per year. From 1973 to 1978, these costs declined, on average, only 4.9 percent annually. . . .
>
> From the mid-1950s to the early 1970s tax credits, accelerated depreciation, and lower corporate income tax rates increased the demand for capital, especially equipment after 1962. Since 1973, however, rising inflation rates generally have retarded the decline in the costs of capital services, thereby depressing the demand for capital assets, especially structures. . . .

> It is no coincidence that capital accumulation was most rapid when capital costs were declining most swiftly, from 1950 until the late 1960s, and that investment has waned recently now that these costs are no longer declining so rapidly.[10]

Another contributing factor to the slower rate of capital accumulation is the decline in the profit rate on capital, making it a less attractive investment. Nordhaus observes that most studies omit this decline in the returns to capital and mix the change in stock with change in utilization:

> The severe recession after 1973, as well as policies which were less pro-investment than in the earlier periods, led to a significantly slower growth in the utilized capital stock. In addition, a point omitted in most studies, the profit rate on capital (and presumably the marginal productivity of capital) has declined in recent years. This would imply that at a given rate of growth of capital the contribution to output would be smaller. There is a serious problem in most of the estimates . . . of the contribution of the capital stock. They compound changes in stock with changes in utilization. The latter appears responsible for most of the contribution of capital to lower growth.[11]

This is substantiated by data from the U.S. Council of Economic Advisors on the rate of return in nonfinancial corporations and the capacity utilization by manufacturing industries. Regarding nonfinancial corporations, the rate of return on depreciable assets before taxes was 16.2 percent in 1965, but half that in 1974 immediately following the severe recession. Regarding manufacturing industries capacity utilization dropped from 89.6 percent in 1965 to 83.8 percent in 1974. In 1980, it dropped further to 79.0 percent. (See table 2.4.)

Statistics from the U.S. Bureau of Economic Analysis show dramatically the impact of the post-1973 recession on business expenditures for new plant and equipment in manufacturing. From 1974 to 1975, expenditures dropped 3.2 percent. The drop is particularly striking compared with changes before and after 1975. The "before" percent change in 1974 over 1970 was three times as much at 9.5 percent, and the "after" percent change in 1976 over 1975 was 9.2 percent. (See table 2.5.)

---

10. Richard W. Kopcke, "Capital Accumulation and Potential Growth," in *The Decline in Productivity Growth* (Boston: the Federal Reserve Bank of Boston, June 1980–1981), 31, 39.

11. William D. Nordhaus, *op. cit.*, 153–154.

Table 2.4

Determinants of Business Fixed Investment: 1960 to 1980

Item	1960	1965	1970	1972	1973	1974	1975	1976	1977	1978	1979	1980
Real investment as percent of real GNP	9.1	10.5	10.5	10.2	11.0	10.9	9.7	9.7	10.3	10.7	11.0	10.7
Capacity utilization rate in mfg.[1]	80.2	(89.6)	79.3	83.5	87.6	(83.8)	72.9	79.5	81.9	84.4	85.7	(79.0)
Nonfinancial corporations:												
Cash-flow[2] as percent of GNP	8.9	10.6	7.8	8.6	8.0	7.0	9.0	9.3	9.6	9.3	8.9	8.6
Rate of return before taxes												
On depreciable assets[3]	11.2	(16.2)	9.5	10.7	10.6	(8.0)	8.8	9.5	10.1	9.8	8.9	7.7
On stockholders' equity[4]	9.9	15.5	8.9	11.3	14.5	13.3	9.5	9.5	11.2	11.9	10.1	9.9
Ratio of market value to replacement cost of net assets[5]	.96	1.25	.86	1.01	.93	.66	.66	.74	.65	.60	.56	(.33)

[1]Federal Reserve Board index. [2]Cash-flow defined as after-tax profits plus capital consumption allowance plus inventory valuation adjustment. [3]Profits plus capital consumption adjustment and inventory valuation adjustment plus net interest paid divided by the stock of depreciable assets valued at current replacement cost. [4]Profits corrected for inflation effects divided by net worth (physical capital component valued at current replacement cost). [5]Equity plus interest-bearing debt divided by current replacement cost of net assets.

Source: U.S. Council of Economic Advisors, Economic Report of the President (U.S. Bureau of Economic Analysis and the Board of Governors of the Federal Reserve System, annual report).

The Sources and Signals of Declining U.S. Productivity

The annual gains since 1975 have been quite healthy. The highest was a 23.8 percent increase in 1979 over 1978. In 1980, however, the annual percent change dropped to 17.4, and in 1981, it dropped further to 9.5.

What has been the effect of business expenditures for new plant and equipment on the capacity of labor to produce? The effect is well-illustrated by data covering the period from 1948 to 1980 from the U.S. Bureau of Labor Statistics and the Bureau of Economic Analysis. The net annual growth in 1972 dollars was 2.6 percent during the 1948–65 period and 2.8 percent during the 1965–73 period, dramatically dropping to 1.2 percent in the period 1973–78. The net annual growth then recovered slightly in 1978–80 to 1.7 percent. Though there was a recovery during 1979–80, to 3.3 percent, the decline in the annual growth is graphic: -1.1% (1975–76), -1.3% (1976–77), -1.7% (1977–78), and -0.1% (1978–79). (See table 2.6.)

0.0 Decline Source: Structural Shifts

It is clear that while expenditures for new plant and equipment have been increasing at a healthy rate, the physical capital per hour worked has been decreasing dramatically, the reason being that hours of work are increasing even more dramatically, as pointed out by Lester C. Thurow:

> Manhours of work were growing 0.4 percent per year from 1948 to 1965, 1.1 percent per year from 1965 to 1972, and 2.1 percent per year from 1972 to 1979. To equip a labor force that was growing more than five times as fast in the last period as in the first period, the fraction of GNP devoted to plant and equipment investment would have had to have risen dramatically. It didn't. . . . Using the three periods from 1948 to 1965, 1966 to 1972, and 1973 to 1979, plant and equipment investment rose from 9.5 percent of the GNP in the first period to 10.3 percent of the GNP in both the next two periods.[12]

Clearly, when the labor force increased 525 percent, if the capital investment grew only 8 percent, the growth rate of physical capital per hour worked would be negative as shown in the annual rates in table 2.6.

Further, the shift in the industrial mix—where labor moved to—accounts for the decline in productivity. Not only is the physical capital per hour worked not growing at the required rate, but labor is moving into industries where the productivity gains are minimal as well. Thurow observes:

12. Lester C. Thurow, "Discussion," in *The Decline*, op. cit., 22.

31

Table 2.5

Business Expenditures for New Plant and Equipment: 1965 to 1983

(in Billions of Dollars, Except Percent.)

Industry	1965	1970	1975	1976	1977	1978	1979	1980	1981	1982	1983
CURRENT DOLLARS											
Total	**70.4**	**105.6**	**157.7**	**171.4**	**198.1**	**231.2**	**270.5**	**295.6**	**321.5**	**316.4**	**310.9**
Manufacturing	25.4	37.0	54.9	59.9	69.2	79.7	98.7	115.8	126.8	119.7	115.9
Durable goods[2]	13.5	19.8	26.3	28.5	34.0	40.4	51.1	58.9	61.8	56.4	54.2
Primary metals	2.6	3.2	5.8	5.8	5.4	5.7	6.8	7.7	8.1	7.5	6.0
Electrical machinery[3]	1.7	3.5	3.1	3.7	4.7	5.7	7.3	9.6	10.3	10.6	11.0
Machinery[4]	2.2	3.8	5.0	5.5	6.6	7.2	10.5	11.6	13.2	12.9	13.1
Transportation equip	3.7	4.6	5.9	6.4	9.1	12.0	15.3	18.2	18.4	15.2	14.6
Percent motor veh	80.2	65.6	56.6	56.7	64.2	60.1	54.2	49.9	54.8	52.2	55.3
Nondurable goods[2]	11.9	17.2	28.6	31.5	35.2	39.3	47.6	56.9	64.9	63.2	61.7
Food and beverage	1.9	3.3	4.0	4.8	5.1	6.0	6.6	7.4	8.2	7.7	7.5
Paper	1.1	1.7	3.0	3.1	3.6	4.0	5.5	6.8	6.7	6.0	6.1
Chemicals	2.8	3.4	7.6	8.1	8.1	8.5	10.8	12.6	13.6	13.3	13.7
Petroleum	3.9	5.2	9.6	10.8	12.7	13.9	16.2	20.7	26.6	26.7	24.6
Nonmanufacturing	45.0	68.6	102.8	111.5	128.9	151.5	171.8	179.8	194.7	196.8	195.0
Mining	1.4	2.0	6.1	7.4	9.2	10.2	11.4	13.5	16.9	15.5	15.5
Transportation	5.5	7.0	8.7	8.9	9.4	10.7	12.3	12.1	12.0	11.9	11.0
Public utilities	6.5	13.0	20.0	22.4	26.8	29.9	34.0	35.4	38.4	42.0	41.0
Electric	4.8	10.6	16.7	18.8	22.2	24.6	27.6	28.1	29.7	33.4	33.1
Gas and other	1.7	2.5	3.3	3.6	4.6	5.3	6.3	7.3	8.7	8.6	7.9
Trade and services	22.1	29.8	46.2	49.3	56.5	68.7	79.3	81.8	86.3	86.9	87.8
Comm. and other[5]	9.6	16.9	21.8	23.5	26.9	32.0	34.8	37.0	41.1	40.5	39.8

change:[6]

Total	7.7[7]	8.5	8.4	8.7	15.6	16.7	17.0	9.3	8.8	-1.6	-1.7
Manufacturing	9.1[7]	7.8	8.2	9.1	15.5	15.2	23.8	17.3	9.5	-5.6	3.2
Nonmanufacturing	6.9[7]	8.8	8.4	8.5	15.6	17.5	13.4	4.7	8.3	1.1	-.9

CONSTANT (1972) DOLLARS[8]

Total	**93.2**	**115.3**	**119.7**	**123.1**	**134.8**	**146.4**	**157.7**	**159.1**	**159.4**	**150.7**	**144.9**
Manufacturing	33.2	40.0	41.2	42.3	46.1	48.8	55.4	60.0	60.8	55.5	53.1
Durable goods	17.3	21.4	20.3	20.7	23.5	25.7	29.7	31.9	31.7	28.5	27.1
Nondurable goods	15.9	18.6	20.9	21.7	22.7	23.1	25.7	28.1	29.1	27.0	26.0
Nonmanufacturing	60.0	75.3	78.5	80.8	88.6	97.6	102.3	99.1	96.7	95.2	91.8
Mining	1.8	2.2	4.1	4.7	5.3	5.1	5.1	5.1	5.4	4.6	4.7
Transportation	6.9	7.5	6.5	6.2	6.3	6.6	6.8	6.0	5.6	5.4	4.8
Public utilities	8.6	14.3	14.2	14.9	16.8	17.5	18.3	17.5	17.3	17.8	17.0
Trade and services	29.7	32.4	36.3	37.6	40.9	46.4	49.4	47.3	47.3	46.3	45.5
Comm. and other[5]	13.1	18.8	17.3	17.5	19.4	22.0	22.7	23.1	23.1	21.1	19.9

Average annual percent change:[6]

Total	6.9[7]	4.2	.8	2.8	9.5	8.7	7.7	.9	.2	-5.5	-3.8
Manufacturing	8.4[7]	3.8	.6	2.7	8.9	5.8	13.6	8.2	1.2	-8.7	-4.3
Nonmanufacturing	6.2[7]	4.7	.8	2.9	9.7	10.2	4.7	-3.1	-.4	-3.5	-3.6

*Excludes expenditures of agricultural business and current account outlays. Based on sample and subject to sampling variability: see text, p. 532. These figures do not agree with the fixed nonresidential investment data mainly because those data exclude investments for equipment, and include investment by farmers and certain outlays to current account. Minus sign (−) indicates decrease. See also *Historical Statistics, Colonial Times to 1970*, series V 306-332.
[1]Estimates based on expected capital expenditures as reported by business in late January and February 1983.
[2]Includes industries not shown separately. [3]Includes equipment. [4]Excludes electrical. [5]Includes construction; social services and membership organizations; and forestry, fisheries, and agricultural services. [6]Change from prior year shown: e.g., 1970 from 1965, except as noted. For explanation of average annual percent change, see Guide to Tabular Presentation.
[7]Change from 1960. [8]Constant-dollar expenditures since 1977 have been reestimated using deflators that reflect revisions of the national income and product accounts released in July 1982.
Source: U.S. Bureau of Economic Analysis, *Survey of Current Business*, monthly.

Table 2.6

Average Annual Growth Rate of Physical Capital per Hour Worked in Private Business: 1943 to 1980

(In percent Rates based on capital valued in constant, 1972, dollars. Excludes nonprofit institutions and government enterprises, Minus sign (−) denotes decrease.)

Item	1948–1965	1965–1973	1973–1978	1978–1980	1973–1974	1974–1975	1975–1976	1976–1977	1977–1978	1978–1979	1979–1980
Total net	2.6	2.8	1.2	1.7	3.3	6.9	−1.1	−1.3	−1.7	−.1	3.3
Equipment	3.7	4.6	2.9	3.1	5.5	8.3	.1	.4	.1	1.6	4.5
Structures	2.3	2.2	.2	1.0	1.8	6.8	−1.9	−2.6	−3.1	−1.0	3.1
Inventories	2.9	3.2	.9	−.1	3.4	3.1	−.5	−.3	−1.0	−.6	.6
Land	1.1	−.4	−.3	−.1	.7	6.7	−1.8	−2.5	−3.9	−2.4	2.2

Source: U.S. Bureau of Labor Statistics and Bureau of Economic Analysis (unpublished data).

From 1948 to 1965, 9.1 million manhours were moved from agriculture to industry. And from 1965 to 1972 another 1.8 billion manhours were released. But in the period from 1972 to 1978, only 0.1 billion manhours were released. . . . [A]griculture entered 1948 with a level of labor productivity just 40 percent of the national average. Every worker moved from agriculture to industry represented a 60 percent gain in productivity . . . [B]ut when this movement stopped, agriculture ceased being a major source of productivity gains. . . .

[S]ervice productivity growth was slow after WWII and by 1978 service productivity was only 62 percent of the national average. Thus a worker moved into services in 1978 represented a 38 percent decline in productivity. And millions of workers were moved. From 1972 to 1978, 5.7 billion manhours or 35 percent of the growth in hours of work went into services. This represented a large reduction in national productivity growth. . . .

Together the agriculture-service effect explains about 22 percent of the decline in productivity growth from the first to the third period.[13]

Using figures from the U.S. Department of Commerce, we can study the actual productivity growth and the effects of the inter-industry shift of resources, e.g., people moving from agriculture to industry, from industry to services, etc. The method is to compute the real product per unit of input with an unchanging composition of labor (or capital) and to compare it with the actual productivity figures. For 1950–66, the weighted productivity growth within major sectors was 2.8 percent, while unweighted average growth was 3.4 percent, the structural effect being the difference between the two, 0.6 percent. Thus, the increase in productivity due to structural shift would be $(0.6/2.8 =)$ 21 percent. During 1966–73, the structural shift effect was 0.3 percent, representing an increase in productivity of $(0.3/2.0 =)$ 15 percent. During 1973–78, the increase due to sectoral shifts was 0.2, representing an increase of $(0.2/1.2 =)$ 17%.

Comparing the last period gain (0.2) with that of the first period (0.6), the gain in productivity due to sectoral shifts is $(0.2 - 0.6 =) -0.4$ percentage point, made up of a (decreased) -0.3 gain in the second period over the first, and a (decreased) -0.1 gain in the third period over the second. However, the productivity of the third period was 56 percent lower at 1.4 percent compared with 3.4 percent in the first period. Structural shifts, therefore, contributed less to productivity

13. *Ibid.*, 23–24.

increases in decidedly declining productivity overall from the first to the third period.

0.0 The Effects of Structural Shifts, by Industry Sector

One of the most likely consequences of robotization would be structural shifts of resources—labor, capital, technology, etc. It is, therefore, instructive to identify the current concentration of structural-shift effects.

Table 2.7 presents the productivity gain in period 3 over period 2 and in period 2 over period 1. We can aggregate these two and derive the productivity gain in period 3 over period 1 and rank-order the industry sector gains. (See table 2.8.)

Thurow found that the agriculture-service effect explains 22 percent of the decline in productivity growth from the first to the third period. He further accounts for another 40 percent as follows:

Table 2.7
Average Annual Rates of Growth in Output per Labor Hour Paid,
U.S. Domestic Business Economy
(In Percent)

Sector	(1) 1950–66	(2) 1966–73	(3) 1973–77	Changes (2)–(1)	(3)–(2)
Agriculture	5.9	5.4	4.2	−0.5	−0.8
Mining	4.5	2.0	−4.4	−2.5	−6.4
Construction	2.9	−2.2	−0.8	−5.1	1.4
Manufacturing					
Durable goods	2.6	1.9	1.3	−0.7	−0.6
Non-durable goods	3.3	3.2	2.6	−0.1	−0.6
Transportation	3.7	2.9	2.1	−0.8	−0.8
Communications	5.3	4.8	6.8	−0.5	2.0
Utilities	6.1	4.0	1.0	−2.1	−3.0
Wholesale trade	3.3	3.9	0.3	0.6	−3.6
Retail trade	2.5	2.2	1.0	−0.3	−1.2
Finance, insurance, etc.	0.8	−0.2	0.6	−1.0	0.8
Services	1.7	1.9	0.4	0.2	−1.5
Weighted productivity growth within major sectors	2.8	2.0	1.2	−0.8	−0.8
Unweighted average growth, Total business economy	3.4	2.3	1.4	−1.1	−0.9
Structural shift effect	0.6	0.3	0.2	−0.3	−0.1

Source: U.S. Department of Commerce, Office of the Chief Economist.

The Sources and Signals of Declining U.S. Productivity

Table 2.8
Contribution to Productivity Growth Decline, by Sector

Sector (in Rank Order 1–12)	Period 3– Period 1	% of Productivity Growth Decline Explained
1. Mining	−8.9	10%
2. Utilities	−5.1	11
3. Construction	−3.7	17
4. Wholesale trade	−3.0	
5. Transportation	−1.6	
6. Retail trade	−1.5	
7. Durable goods	−1.3	
8. Agriculture	−1.3	22*
9. Services	−1.3	
10. Nondurable goods	−0.7	
11. Finance, insurance, etc.	−0.2	
12. Communications	+1.5	
Weighted productivity growth within major sectors	−1.6	
Unweighted average growth, total business economy	−2.0	
Structural shift effect	−0.4	

*Agriculture-service effect
Basic Source: Table 2.7.

> Construction productivity has now been falling for 15 years and accounts for about 17 percent of the decline in national productivity. . . . The decline in extractive industries explains about 10 percent of the national decline in productivity. . . .
>
> [T]here has been a sharp decline in the rate of growth of productivity in the utility industry.[14]

While Thurow does not give the percentage of productivity growth explained by the utility industry, we note that the utilities are placed between mining and construction in rank order of growth decline. If we interpolate on the basis of the growth decline, we can attribute 11 percent of productivity growth decline to utilities.

14. Lester C. Thurow, *op. cit.*, 24.

The four industry sectors together account for 60% of the productivity growth decline—agriculture-services (22%), construction (17%), utilities (11%), and mining (10%).

0.0 Decline Source: Research

On the basis of the several studies cited in table 2.3, research is one-third as important as capital and sectoral shifts combined, but there is a strong divergence of expert opinion on the role of research and development in explaining the U.S. decline in productivity growth.

Kendrick says:

> Over the long run, by far the most important source of productivity growth is advances in technological knowledge applied to productive processes and instruments. Cost-reducing innovations in the ways and means of production are particularly important, but even the development of new products for sale to consumers and governments as well as to producers affect productivity indirectly through the learning-curve effect although GNP estimates do not adequately reflect the gains in welfare due to new and improved goods and services.[15]

Robert Solow questions the concept of a "stock" of research and development:

> Gilriches and Kendrick, and no doubt others, proceed by calculating a "stock" of R & D, a cumulation of real current expenditures less some sort of depreciation of old knowledge. Then they treat this stock as a factor of production in the ordinary growth-accounting framework. Gilrich gives it an output-elasticity of 0.06; Kendrick appears to use a larger estimate, say 0.10 or 0.12. It is a reasonable approach and I do not know a better one. But it is hardly self-evident that productively useful knowledge behaves like a stock, is added to by R & D spending with some specifiable lag after the money is spent, and has a marginal product like a conventional input.[16]

Thurow expresses skepticism for another reason:

15. John W. Kendrick, *op. cit.*, 8.
16. Robert M. Solow, "Discussion," in *The Decline, op. cit,* 176.

> Kendrick points to declining expenditures on R & D as
> one of the sources of the productivity decline. I am skep-
> tical because of the timing. R & D expenditures peaked
> as a percent of GNP in the mid-1960s, but they did not
> really start to decline until the early 1970s. The produc-
> tivity decline started much earlier and there must be a
> substantial time lag between R & D expenditures and
> measured productivity.[17]

Solow's comments on Kendrick appear to express disagreement on the process rather than the concept itself. It appears that Solow agrees with Kendrick that
research and development does indeed stimulate productivity, but wonders exactly
how it is brought about. Thurow is skeptical about the role of research and development on productivity because productivity growth decline started as long as it
did before growth decline in research and development expenditures.

The National Science Foundation provides data on research and development
outlays, which cover basic research, applied research, and development. We see
that research and development increased from 2.67 percent of the GNP in 1960 to
2.90 percent in 1965, but then declined to 2.65 percent in 1970 and further to
2.29 percent in 1979.

Perhaps more significant than the fraction of GNP spent on research and development, and more significant than the size of research and development, is its
distribution. Defense received more than half the research and development outlay
in 1960, but only less than a quarter in 1979. The share of research and development spent on space rose from 3 percent in 1960 to 7 percent in 1979; as did the
share of Federal nondefense and nonspace research and development from 9 percent to 18 percent. The nonfederal nondefense, nonspace share rose from a third
in 1960 to more than half in 1979. (See table 2.9.)

The nonfederal percentage of research and development performance rose by
43 percent from 37 percent in 1965 to 53 percent in 1979. During this period,
the research and development outlay by industry rose from $3.2 billion to $9.4
billion.

0.0 Why is U.S. Productivity Declining?

We have seen that by any account, the increase in labor productivity in the U.S.
has been deceasing.

0.0 The Falling Star

Three primary sources of this decline are capital, sectoral shifts, and research.

To increase the output per hour, capital equipment which multiplies the labor

---

17. Lester C. Thurow, *op. cit.*, 24.

Table 2.9
Research and Development Outlays by Sectors

	1960	1965	1970	1979
Total (Bil. current $)	13.5	20.0	25.9	54.3
GNP (Bil. current $)	506.0	688.1	982.4	2368.9
Research and development as percentage of GNP	2.67%	2.90%	2.64%	2.29%

Basic Research, Applied Research and Development
(Percentage of Total Research and Development Outlays by Sectors)

	1960	1965	1970	1979
Defense	52%	33%	33%	24%
Space	3	21	10	7
Nondefense, nonspace—federal	9	11	14	18
Nondefense, nonspace—Nonfederal	36	35	43	51

Basic Research, Applied Research and Development
(In Billions)

	1965	1970	1979
Defense	6.4	10.3	25.2
Space	0.4	0.6	1.9
Nondefense—federal	0.6	0.9	1.8
Total Federal research and development	7.4	11.8	28.9
Total research and development	20.0	25.9	54.3
Nonfederal research and development as percentage of total research development	37%	45%	53%

Source: National Science Foundation, *National Patterns of Science and Technology Resources, 1980* (Washington, D.C.: 1981).

input is critical. While the 2.1 percent increase during 1972–79 in labor force was 525 percent of the corresponding figure of 0.4 percent increase in 1948–65, capital investment grew only 8 percent from 9.5 percent of the GNP in the 1948–65 period to 10.3 percent in the period from 1973–79. Why? "Economic" regulation, as distinguished from "social" regulation pertaining to pollution and occupational safety. It has not been sufficiently attractive for people to save money and deposit it in the banks; it has not been sufficiently attractive for the banks to lend money to business; and it has not been sufficiently attractive for business to invest in plant and equipment.

If investment in plant and equipment in conventional machinery itself is hard to come by, what will happen to unconventional machinery of the type demanded by the revolution in robotics? Not only are they unconventional, but they are much more expensive as well.

We may call the "economic regulation" elements by different names, such as tax incentives, depreciation and amortization, saving stimulus, but they are all regulations which impact upon the increase in the physical stock of equipment that multiplies human production effort (capital), and the intellectual stock of knowledge that multiplies human production effort (research and development).

0.0 The Rising Sun

The decline and fall in productivity in the U.S. is a study in contrast with the rise of productivity in Japan. Frank Gibney says in his book *Miracle by Design*:

> The comparison is so dismal as to be almost a caricature. Where the Japanese save we spend. Where Japanese companies invest our companies divest. They conserve, we conspicuously consume. Where the Japanese find new jobs for people in depressed industries, we fire them and close the plants. Where the Japanese tighten the standards of their education, we loosen ours, with predictable effect. Where the Japanese keep zealously expanding their total product, we fight over what we have left.
>
> Japanese capitalism is a study in growth, growth agreed upon and growth pursued. The American economy seems by contrast a vast Brownian movement. All its particles are wildly in motion—suing, arguing, firing, and denouncing—but going nowhere.[18]

The key to the U.S.-Japanese productivity is Gibney's observation of "growth agreed upon and growth pursued." Agreement is found among all the main elements—capital, labor, government, and university. The "agreement" is

18. Frank Gibney, *Miracle by Design*, (New York: Times Books, 1982), 214.

41

practically as binding as an iron-clad contract signed in blood. The reason is institutional. It is the ethos of their existence. It is as fundamental as their personal identity. Unwritten, ambiguous "understanding" among the individual members of a family, corporation, clique, or some other group prescribes and determines collective behavior. In this context, it is indeed meaningful to say "growth agreed upon." Given this kind of agreement, "growth pursued" is much more than the increase in one's paycheck or the increase in sales and profits of one's company. It is a point of national honor.

These are concepts and protocols which appear to the American mind at best strange and at worst unworkable abstractions. They undeniably work, however, and work well for Japan. Since the question "robotize or retrogress" is no longer an exercise in abstraction, but a matter of survival itself, we must examine what makes Japanese capital, labor, government, and university work hand in glove to agree upon and pursue, wholeheartedly, economic growth. To this we turn next.

3

The Product, Process, and Premises of Rising Japanese Productivity

OVERVIEW: We can group the factors of Japanese productivity according to the interrelated and integral elements of a triad—institutional, investmental, and inventional/innovational. Productivity, to the Japanese, is "a way of practising virtue." The Confucian preeminence of relationships among individuals binds individuals to each other and to groups through obligations of specified order.

Among the "top down" obligations are respect for seniority, provision for retirement, and lifetime employment. Among the "bottom up" obligations are *On*, obligations passively incurred, and *Gimmu* and *Giri*, active repayments of *On*. The top down and bottom up obligations institutionalize the interdependence of employee and employer. From the institutional commitment come the investmental factors—the administrative guidance, the beneficial guidance of the banks, and the special stimulus to companies to invest in research and development. Given the institutional and investmental factors, the Japanese pursue the inventional and the innovational by seeking to "gain the whole world (market) first." The strategy to gain the whole world market is a strategy of end-running, never launching a frontal attack, but concentrating heavily on areas where most advantage can be gained.

In the 1950s and 1960s, they concentrated on manufacturing engineering. Later, they shifted to quality control and product design. Currently, they focus on basic research and direct marketing. The Japanese implement extremely detailed and pragmatic planning for each upcoming six-month period by tracking every variance meticulously and striving to do better during the following period.

Clearly, in many respects, the triad of Japanese productivity is unique. If the factors that make the Japanese triad work can be identified, they can be studied to determine ways of emulating them and adapting them to improve American productivity.

Robotics/Artificial Intelligence/Productivity

0.0 Samurai and Other Schools of Management

The obvious success of Japanese industry to produce quality goods at competitive prices in recent years has naturally led to a search for the secrets of their success. A predominant theme of the books and articles on the subject is what may be called the Samurai School of Management. Its feudalistic underpinnings would probably fail to inspire American capitalists to immediately take up the sword of the samurai school of management, but some understanding of the roots of the Japanese way would shed light on what is working so well there.

Frank Gibney, who in ten years multiplied *Encyclopedia Britannica*'s Japanese business seventeen-fold to $100 million, using Japan's own methods, traces the Japanese work ethic to Confucian capitalism, whose articulate spokesman in the nineteenth century was Shibusawa Eiichi. Shibusawa even coined the modern word for businessman, *jitsugyoka*, which Gibney translates as "a practical man of affairs." Shibusawa combined ethics with mathematics. His aim, as Gibney puts it in his book, *Miracle by Design*, was "building modern enterprise on the abacus and the Confucian *Analects*."[1]

The Confucian theme of primacy of the five basic relationships among father and son, ruler and official, husband and wife, brother and sister, and friend and friend is echoed in William Ouchi's Theory Z which accents trust, subtlety, and intimacy. In his book, *Theory Z*, Ouchi says: "More than hierarchical control, pay, or promotion, it is our group memberships that influence our behavior. . . . we must adapt only those aspects of Japanese management that suit our needs . . . such as trust, loyalty to a firm, and commitment to a job over most of one's productive years . . . are the foundation of Theory Z."[2]

Ouchi served as a collaborator during the recruitment phase of a McKinsey & Company study of Japanese and American firms by his Stanford University colleagues Richard T. Pascale and others. The study led to the formulation of the 7 S Framework in the book, *The Art of Japanese Management*, by Richard T. Pascale and Anthony G. Athos. They point out that the art of Japanese Management is different from the American because, by and large, the American management tends to emphasize the hard Ss of strategy, structure and systems which produce "an arid world in which nothing is alive," while the Japanese concentrates on the four soft Ss of staff, skills, style, and superordinate goals, "through which an organization is often given its life."[3]

Interestingly enough, in his book *The Mind of the Strategist*, a Japanese alumnus of MIT who is now managing director of the Tokyo office of McKinsey

1. Frank Gibney, *Miracle by Design,* (New York, N.Y.: Times Books, 1982), p. 30. Excerpts by permission of Random House.
2. William G. Ouchi, *Theory Z*, (Reading, Mass.: Addison-Wesley, 1981), pp. 25, 22. Excerpts by permission.
3. Richard T. Pascale and Anthony G. Athos, *The Art of Japanese Management* (New York: N.Y.: Warner, 1981), pp. 123 ff, 1.

44

& Company, Kenichi Ohmae, explains the art of Japanese firms not in terms of the soft Ss that the McKinsey framework associates specifically with Japan, but of the very first hard S, strategy. He defines strategy "as the way in which a corporation endeavors to differentiate itself positively from its competitors, using its relative corporate strengths to better satisfy customer needs."[4]

We will examine the insights of these students of Japanese management as well as others and attempt an operational integration of the successful precepts. We start with the triad of productivity as an integrating analytical framework.

0.0 The Triad of Productivity

To meet the Japanese challenge of productivity, a challenge made even more acute by the Japanese commitment to robotize, it is critical to understand the forces that have catapulted the militarily vanquished in 1945 to the economically victorious in the 1980s. Analysis reveals three groups of factors integral to the Japanese success, factors as interrelated as the elements of a musical triad. We label these *institutional, investmental,* and *inventional/innovational.* Whether we emulate, adapt, and excel in the Japanese approach that incorporates these factors or fail to learn and use the factors that work so well for the Japanese is critical to the question we raised in chapter 1 — "robotize or retrogress?"

By and large, we may find the institutional factors unique to the Japanese culture and, therefore, we may find it hard enough to emulate them, let alone to excel in them. And a study of the investmental factors highlights an active cooperation with industry by government and banks in Japan that has no parallel in the United States. It may be possible, however, to find in the inventional/innovational factors parallel sources of strength unique to our own culture.

In looking at the inventional/innovational factors, we focus attention on the concentration by the Japanese on first finding, then serving, the market. Invention, the creation of something new, is neither necessary nor sufficient for innovation — the creation of something perceived to be novel, the introduction of something to some degree different. By concentrating on innovation instead of invention, the Japanese marketeer first gains the market and then expands it by skillful service.

0.0 The Institutional Factors of Japanese Success

What makes a man want to do what he does is indeed the most significant key to understanding him. What makes him tick? What makes the Japanese tick? American businessman Frank Gibney, having succeeded in Japan using Japanese methods, has some claim to expertise in explaining the Japanese business psyche to the American.

---

4. Kenichi Ohmae, *The Mind of the Strategist,* (New York, N.Y.: McGraw-Hill, 1982), p. 92.

Robotics/Artificial Intelligence/Productivity

"Productivity is a Way of Practising Virtue"[1]

In his book, *Miracle by Design*, Gibney credits Shibusawa Eiichi with bringing respectability to the businessman in the third quarter of the nineteenth century. No mean businessman himself, Shibusawa founded Japan's largest bank, Dai Ichi Kangyo, helped organize some five hundred companies, and helped set up the Ministry of Finance in the early Meiji government. Writes Gibney:

> As a reaction to the family-based *zaibatsu* [money cliques], he introduced the joint-stock company to Japan. This, he felt, was a more community-type company, which could marshall existing capabilities to better advantage.
>
> But Shibusawa's influence stretched beyond his actual business and financial achievements. He was the conscience of Japanese business. It was he who developed the work ethic of Suzuki Shōzan [seventeenth century Zen master] and Ishida Baigan [eighteenth century Confucian] and applied it to the world of developing Japanese industry. Where people like Iwasaki [Yataro, the founder of the Mitsubishi group] sought to maximize profit, for Shibusawa, profit was not the primary motive. He believed that more profit would accrue in the end to the honest businessman who planned wisely and justly and took the long view of development, with the interests of the country as well as the company in mind. . . .
>
> As he wrote later in life, explaining why Confucian virtue and business should go together: "Morality and economy were meant to walk hand in hand. But as humanity has been prone to seek gain, often forgetting righteousness, the ancient sage, anxious to remedy this abuse, zealously advocated morality on the one hand and on the other, warned people of profit unlawfully obtained. Later scholars misunderstood [Confucius'] true idea. . . . They forgot that productivity is a way of practising virtue."

Productivity, it is seen, is not a means to plenty, but a means to piety. Conversely, if one is not productive, he fails in his divine obligations. To do one's job well is, in a sense, to achieve self-realization.

"Gain salvation through doing work well"[3]

Explicating the moral rationale for economic work, Yamamoto Shichihei, whom Gibney describes as "one of Japan's most prominent social commentators and essayists," says:

> We still believe in the teaching of Suzuki Shōzan that by doing our daily work we are doing the Buddha's work as well. Thus when a manufacturer makes good products he is showing one face of the Buddha *(ichibutsu no bunshin)* to bring profit to the world. When a salesman makes his rounds, he is on a pilgrimage. Each in his own task can gain salvation through doing work well—as long as he keeps away from the "three poisons" of greed, anger, and idle complaint. . . .
>
> No, we work hard because we believe in it. To do a good job and help one's company grow and prosper—that for so many of us in Japan is our *ikigai*—what makes our life worth living.

The obligation—Bottom up

Why is it that helping one's company grow and prosper is so vital to the Japanese? The answer is in the Confucian preeminence of relationships among individuals, almost to the point of ignoring individual rights and privileges altogether. Chitoshi Yanaga, in *Big Business in Japanese Politics*, describes a powerful emotional force called *on*:

> Perhaps no other emotional force in Japanese society is as powerful and propulsive as that which inheres in the sense of obligations and indebtedness which is known as *on*. From the cradle to the grave, this emotional force, in all its variations and manifestations, propels the individual in his actions. . . . Obligations bind the individual to individuals and groups in a tight web of relationships which ignore individual rights and desires. What exists is a hierarchy of obligations beginning at the top with loyalty to the emperor and the state, to society at large, to the family as manifested in filial piety, to one's superiors, teachers, friends, in-laws, and even to subordinates and servants.[5]

Earlier, we referred to five basic relationships: father and son, ruler and official, husband and wife, brother and sister, and friend and friend. Each one of these relationships not only engenders binding obligations, but also sets the stage for other binding obligations of higher and lower order. In a listing by Yanaga, we recognize a descending order of obligations: emperor and state, society at large,

5. Chitoshi Yanaga, *Big Business in Japanese Politics* (New Haven, Conn.: Yale University Press, 1968), 24.

family, superiors, teachers, friends, in-laws, subordinates and servants. In *The Chrysanthemum and the Sword*, Ruth Benedict shows the obligational hierarchy and its required reciprocity in the table of obligations and reciprocals.

Schematic Table of Japanese Obligations and Their Reciprocals

I. *On:* obligations passively incurred. One "receives an *on*"; one "wears an *on*," i.e., *on* are obligations from the point of view of the passive recipient.

ko on. On received from the Emperor.

oya on. On received from parents.

nushi no on. On received from one's lord.

shi no on. On received from one's teacher.

on received in all contacts in the course of one's life.

NOTE: All these persons from whom one receives *on* become one's *on jun, "on* man."

II. Reciprocals of *on.* One "pays" these debts, one "returns these obligations" to the *on* man, i.e., these are obligations regarded from the point of view of active repayment.

A. *Gimu.* The fullest repayment of these obligations is still no more than partial and there is no time limit.

chu. Duty to the Emperor, the law, Japan.

ko. Duty to parents and ancestors (by implication, to descendants).

nimmu. Duty to one's work.

B. *Giri.* These debts are regarded as having to be repaid with mathematical equivalence to the favor received and there are time limits.

1. *Giri*-to-the-world.

Duties to liege lord.

Duties to affinal family.

Duties to non-related persons due to *on* received, e.g., on a gift of money, on a favor, on work contributed (as a "work party").

Duties to persons not sufficiently closely related (aunts, uncles, nephews, nieces) due to *on* received not from them but from common ancestors.

2. *Giri*-to-one's-name. This is a Japanese version of *die Ehre.*

One's duty to "clear" one's reputation of insult or imputation of failure, i.e., the duty of feuding or vendetta. (N.B. This evening of scores is not reckoned as aggression.)

One's duty to admit no (professional) failure or ignorance.

One's duty to fulfill the Japanese proprieties, e.g., observing all respect behavior, not living above one's station in life, curbing all displays of emotion on inappropriate occasions, etc.[6]

Notice that the "Duty to one's work" engenders a *gimu,* which has to be repaid actively. The fullest repayment of these obligations is still no more than partial and there is no time limit. In other words, one's obligation to work, i.e., to the company, is a continuing obligation which must be actively repaid. Little wonder, "idling" is always used as a pejorative in Japanese. An idler is not paying his service to the Buddha, he is defaulting on his *gimu* to work.

The Obligation—Top Down (i) Employer's Respect for Worker's Seniority
The table of obligations and reciprocals is not one-sided. Recall that Yanaga's listing includes one's obligations to subordinates and servants. The company has certain obligations to the employees, even as they have to repay the *gimu.*

If we take the reference to physical endowments to mean the contribution to society, "Age before beauty" is a theme of Japanese life. Ouchi narrates an anecdote about a new factory in Japan owned and operated by an American electronics company:

In the final assembly area of their new plant long lines of young Japanese women wired together electronic products on a piece-rate system: the more you wired, the more you got paid. About two months after opening, the head foreladies approached the plant manager. "Honorable plant manager," they said humbly as they bowed, "we are embarrassed to be so forward, but we must speak to you because all of the girls have threatened to quit work this Friday." (To have this happen, of course, would be a great disaster for all concerned.) "Why," they wanted to know, "can't our plant have the same compensation system as other Japanese companies? *When you hire a new girl, her starting wage should be fixed by her age.* An eighteen-year-old should be paid more than a sixteen-

6. Ruth Benedict, *The Chrysanthemum and the Sword* (Boston: Houghton Mifflin, 1946), 116.

year-old. *Every year on her birthday, she should receive an automatic increase in pay. The idea that any one of us can be more productive than another must be wrong, because none of us in final assembly could make a thing unless all of the other people in the plant had done their jobs right first.* To single one person out as being more productive is wrong and is also personally humiliating to us." The company changed its compensation to the Japanese model.[7] (emphasis added)

What the incident at the factory showed is the ingrained nature of the Confucian hierarchical order, with the father at the top and the youngest daughter at the bottom of a scale of precedence, strictly observed. The hierarchy is strictly respected in the Japanese corporate professional ranks as well:

> Part of the complex of intertwined features of the Japanese organization are approaches to evaluation and promotion. Imagine a young man named Sugao, a graduate from the University of Tokyo who has accepted a position at the fictitious Mitsubeni Bank, one of the major banks. For ten years, Sugao will receive exactly the same increases in pay and exactly the same promotions as the other fifteen young men who entered with him. Only after ten years will anyone make a formal evaluation of Sugao or his peers; not until then will one person receive a larger promotion than another.[8]

Here again, we find that chronological age, which is the arbiter of protocol at home, is also the arbiter of protocol in the factories and in executive offices. The basic idea is that one's contribution improves with age (up to a point), and that it is unproductive to expect significant contributions disproportionate with age.

This provides the basis for the seniority system, in which age is definitely a significant factor, although not the only significant factor. The premise seems to be that unless proved less productive, the longer the time the person spends with the company, the higher his contribution to the company, and therefore the higher his rewards.

7. William G. Ouchi, *Theory Z*, (New York: Avon, 1981), 41.
8. Ouchi, *op. cit.*, 22.

The Obligation—Top Down (ii) Provision for Retirement
The obligations of the company to its employees is observed not only during the productive periods of one's employment, but also in the handling of one's retirement. Gibney writes that:

> In Japan the government payments for social security are not great. Beginning at the age of sixty, each Japanese retiree receives roughly only between $6,000 and $8,000 annually, with certain medical benefits appended. Retirement ages in companies, however, which were originally set at fifty-five, have been moved up to sixty, and may possibly go higher. But in the Japanese business tradition, retired employees are generally hired again on a contract basis—and at reduced salaries—or given opportunities with subsidiary companies.[9]

The Obligation—Top Down (iii) Commitment to Suppliers and Subsidiaries
The "subsidiary companies" could include suppliers to the company one is retiring from. The life-long employment security is enjoyed by less than 30 percent of the Japanese workforce, which work for the 34,000 major companies employing more than 300 people. The remaining 70 percent of all Japanese production workers work in the small-industry sector. A large segment of small industry depends on the major companies for their survival. Matsushita Electric Industrial Co., Ltd., for example, has more than 500 such affiliates supporting its operations. The binding obligations of the major companies to their suppliers, members of the small-industry sector, are taken quite seriously.

> The relationship between satellites and major firms in Japan constitutes a bilateral monopoly, in which the satellite has only one customer for its product and the major firm has only one supplier for each of its inputs. Such a relationship can easily degenerate into one of mutual distrust and bickering where each side accuses the other of taking advantage and demands the protection of extensive contractual specificity and close auditing of performance. . . .
> In a Japanese auto plant, however, the relationship between the major auto firm and its satellite suppliers is one of total cooperation. So close is the working relationship

9. Gibney, *op. cit.*, 216.

that the supplier will deliver parts right to work stations on the assembly line. Furthermore, the supplier will gladly deliver only a small amount of the part, perhaps enough to last for only three hours of production. Thus, an assembly plant does not have to maintain a large inventory of parts and saves valuable space.[10]

The Obligation—Top Down (iv) Lifetime Employment
The dominant relationship of major companies with their satellite firms simultaneously assures steady supply of parts, with little or no inventory space taken up at the major manufacturer's facility, and the virtual elimination of competition from them. This relationship is one of the three main reasons why Ouchi thinks lifetime employment is possible in Japan:

> Lifetime employment is possible only as a consequence of a unique social and economic structure *not replicated in the United States*. Consider three major factors: first, every major firm in Japan pays all of its employees a large share of their compensation in the form of a bonus, typically paid each six months. The bonus each year adds up to five or six months worth of salary for each employee, and since all receive the same fraction of their salary, the amount is not contingent on individual performance, but only on the performance of the firm. . . . The lifetime employment system allows a firm to pay a small bonus in a bad year or even defer the payment of the entire bonus to a later year. Thus *a firm can cut its payroll by perhaps thirty percent without laying anyone off*. . . .
>
> Second, every major firm in Japan has a large category of temporary employees who are mostly women. . . . Although they may work for the next twenty years, women are considered temporary employees and are immediately laid off in slack periods. . . .
>
> Third, consider for a moment the problem facing the satellite firms. . . . More typically, small firms cannot obtain licenses to import the raw materials necessary to manufacture in major industries, and they exist instead as suppliers to the major firms. The major firms *contract out to them those services most susceptible to fluctuation*, with the result that during a recession, these small firms will sharply contract or go out of business.

10. Ouchi, *op. cit.*, 16, 17.

The combination of bonus payments, temporary employment, and satellite firms provides a substantial buffer
against uncertainty that makes a stable, lifetime employment a reality for the male employees of major firms.[11]
(emphasis added)

The Ritual of Acculturation

Clearly, the understanding and acceptance of myriad sets of obligations takes a
continuous process of internalization. From the first time a child is taught to observe the proper deference in manner and speech in talking with his immediate el-
der, the process starts. When the child goes to school, the process continues. At
every turn, the child picks up nuances of deference, decorum, and diplomacy.
When a person enters a company, particularly on a basis of lifetime employment,
he continues to learn new behavior patterns through a process of acculturation.
Confucius said:

Govern the people by regulation, keep order among you
by chastisements and they will flee from you and lose all
self-respect. Govern them by moral force, *keep order
among them by ritual* and they will keep their self-respect
and come to you of their own accord.[12] (emphasis
added)

Ritual is of a piece with the religious aura of work, as illustrated in Shibusawa's observation that "productivity is a way of practising virtue," and Yamamoto's adage, "Gain salvation through doing work well." According to
Yamamoto, "Our worldly activity and achievement become part of a religious exercise." The religious exercise in the corporate setting includes the lusty singing
aloud of the company song by grown men and women. Matsushita, today the
largest manufacturer of electrical appliances in the world, was the first Japanese
company to institute a company song. When IBM's Thomas Watson, Sr., visited
Matsushita in the 1930s, it made a profound impression on him. By the 1940s,
IBM employees were singing their company song in America. The two songs[13]
are as follows:

Matsushita Workers' Song

For the building of new Japan,
Let us put our strength and mind together,

---

11. Ouchi, *op. cit.*, 20, 21, 22.
12. Confucius, *The Analects*, Book 2.3.
13. Quoted in Herman Kahn, *The Emerging Japanese Superstate* (Englewood Cliffs, N.J.: Prentice Hall, 1970), 110, 112.

Robotics/Artificial Intelligence/Productivity

Doing our best to promote production,
Sending our goods to the people of the world,
Endlessly and continuously,
Like a water gushing forth from a fountain.
Grow, industry, grow, grow, grow!
Harmony and sincerity!
Matsushita Electric!

IBM Anthem Chorus

Ever Onward! Ever Onward!
That's the spirit that has brought us fame!
We're big, but bigger we will be.
We can't fail for all can see
That to serve humanity has been our aim!
Our products are known in every zone.
Our reputation sparkles like a gem!
We've fought our way through—and new
Fields we're sure to conquer too
For the Ever Onward IBM.

The Investmental Factors of Japanese Success

From the brief survey of institutional factors of Japanese success, we turn to the investmental factors.

0.0 Close Cooperative Relationship Between Government and Business"[14]

Without the close-knit relationship between government and business obtained in Japan, it is hardly conceivable that Japan would have been able to develop as she has done. Historical necessity created the alliance.

In 1868, with western powers at the gates demanding free trade, united samurai defeated the last of the Tokugawa shoguns who had ruled Japan for 250 years, and restored the Meiji emperor to power. To make Japan strong to resist western colonization, indigenous forms of capitalism had to be evolved:

Since no modern industry existed, the government itself
had to start up the new factories and sell them off to private entrepreneurs or investors in the new joint-stock

14. Gibney, *op. cit.*, 27.

companies. So emerged the close cooperation between the government and business. Almost no capital was available for starting major modern industries. Hence the government had to set up a national taxation system, then establish a national bank, the Bank of Japan, to supply money and control it, through new city banks.[15]

In 1874, the government essentially set up in business the Mitsubishi group, the largest general trading company group, by giving its founder, Iwasaki Yataro, thirteen ships to ferry troops in the Formosan expedition. Similarly, Fujitsu, the leading computer group, got its start in 1877 when Furukawa Ichibei received a government charter to develop the copper mines at Ashio.

0.0 Administrative Guidance (gyōsei shidō)

The paternalistic governmental control over Japanese industry is exercised today via the twenty thousand Kasumigaseki bureaucrats in the Finance Ministry and Economic Planning Board, the Ministry of International Trade and Industry (MITI), and the Foreign Office. As arbiters of the destiny of Japan itself, they command very high prestige indeed:

> Far from his opposite number in the American civil service, the Japanese bureaucrat sits at the top of the heap in the society. Competition is intense for the demanding tests for "high level public officials." In 1980 some 50,000 young graduates applied; fewer than 1,300 were accepted. . . .
>
> Their most effective tool is administrative guidance (gyōsei shidō), by which they set goals for various sectors of the economy, control inflation and the money supply, identify growing industries and help them expand, adjust exports in response to protests from foreign countries (and pressure from the Foreign Office), and try to divert the energies of companies in depressed industries toward diversification into other fields.
>
> The guidance may be positive, as with the computer industry, for example. As early as the *1950s, MITI had started to encourage the Japanese computer industry through a variety of government subsidies and joint government and industry research projects.* . . .
>
> MITI, for example, now calls Japanese carmakers in for

15. *Ibid.*

> administrative guidance about the number of cars they
> should export. It is MITI that for some time has been
> pressuring Toyota, Nissan, and others to *put up plants in
> the United States*. The government is forever adjusting
> supply and demand in the steel industry, with monthly
> meetings held to pool information about the projected
> needs of big steel users. Based on these, MITI issues a
> quarterly "Outlook for Steel Supply and Demand" as a
> guideline for steelmakers to plan production.[16] (emphasis
> added)

The administrative guidance by MITI is directed toward the growth-industry of the day. In the 1960s, it was heavy industries. In the 1970s, it was the
knowledge-intensive industries. In the 1980s, it appears to be Very Large Scale
Integrated (VLSI) circuits. But growth-industry policy is not the only concern of
MITI:

> The growth-industry policy is but one, albeit important,
> aspect of the broader industrial policy of the MITI. Addi-
> tionally, the ministry takes measures to promote rational-
> ization and reorganization of industry. . . . Whether the
> task is one of rationalization, reorganization, or protec-
> tion, the predictable MITI prescription has been to limit
> competition and promote bigness.
>
> The ministry's administrative guidance techniques play an
> important role here. The cliental relations, which each
> sector of the economy has with the industrial (vertical)
> bureau of the ministry and the consultative mechanisms
> of advisory councils and trade associations, are very use-
> ful to the MITI in carrying out its guidance activities. The
> ministry's mediation of public and private loans is also an
> important instrument of persuasion.[17]

0.0 At the Center of Each Group—the Bank

The deliberately anonymous bureaucrat, whose pay is modest and whose promotion is slow, must wait for entry into the private sector to earn a significant salary.
Once he achieves this, however, as chairman, vice-chairman, or president of a
private company, he can exercise decisive influence on the professional careers of
others.

---

16. Gibney, *op. cit.*, 135, 136, 137.
17. Kangi Haitani, *The Japanese Economic System* (Lexington, Mass.: Lexington Books, 1976), 135–136.

Ouchi reports that MITI bureaucrats wield great power not only in providing administrative guidance to corporations, but also in providing retirement placement to corporate employees. The plum assignments are in the commercial banks. Why? Because they are at the center of the corporate group, just as they were in the family-based *zaibatsu*.[18]

In the late nineteenth century, during the Meiji restoration period, the new large enterprises were modelled after the family merchant houses of the preceding Tokugawa period:

> There were four major *zaibatsu* (money clique) enter-
> prises—Mitsui, Mitsubishi, Sumitomo, and Yasuda—
> conglomerates with interests in almost every branch of in-
> dustry. Mitsui's prewar interests, for example, included
> mining, shipping, oil, cement, automobiles (Toyota was
> originally a Mitsui company), chemicals, precision ma-
> chinery, textiles, flour, real estate, and of course, at the
> center of the group's activities, the Mitsui Bank . . .
> The House of Mitsui . . . helped finance the emperor's
> forces. . . .[19]

While the *zaibatsu* were broken up, and the leadership retired, by the MacArthur trustbusters, Ouchi observes that the relationships continue:

> Although the *zaibatsu* were legally dissolved after the
> war, the relationships continue. This is due largely to the
> dependence of all companies in a closely-knit network of
> allied banks for their financing. The banks see to it that
> no one firm takes advantage of a trading partner, and so
> the spirit of *Zaibatsu* survives. . . .
>
> The relationship between the *Zaibatsu* and the system of
> lifetime employment is an intimate one. . . . The bank, at
> the center of each group, has even greater power to place
> retiring employees into any number of major firms or into
> their satellites. Consequently a career with a major bank
> is highly prized in Japan. Perhaps the ultimate goal, how-
> ever, is a career at MITI, the Ministry of International
> Trade and Industry, which regulates all business in Japan.
> Only MITI can place retirees in a commercial bank.[20]

18. Ouchi, *op. cit.*
19. Gibney, *op. cit.*, 28.
20. Ouchi, *op. cit.*, 17, 18.

Robotics/Artificial Intelligence/Productivity

The *zaibatsu* centrality of the bank continues today in the general trading company *(sogo shosha)*. While the term originally referred to the comprehensive (sogo) list of merchandise it handled, today it refers to the conglomerates, the top ten handling among themselves merchandise worth more than two-fifths of the national budget. The bank connection is evident in the following listing of the ten largest general trading companies:[21]

General Trading Company		Banking Affiliate
Mitsubishi Corp.	$31.4 billion	Mitsubishi
Mitsui & Co.	28.8	Mitsui
Marubeni Corp.	18.5	Fuji
C. Itho & Co.	17.4	Sumitomo and Dai-ichi Kangyo
Sumimoto Shoji	17.1	Sumitomo
Nisso-iwai Co.	13.4	Sanwa and Dai-ichi Kangyo
Tomen	8.1	Mitsui and Tokai
Kanematsu-Gosho	7.7	Dai-ichi Kangyo
Ataka & Co.	7.0	Sumitomo
Nichimen	6.9	Sanwa

The bank is decidedly central in Japanese industry. It usually makes low-interest loans in the short-term, which are rolled over into higher-rate, long-term loans, encouraging Japanese firms to assume debts four and five times as much as equity. The bank, which owns some five percent of voting stock in client firms, does take a *long-term* view of returns:

> Where American management typically raises its money by issuing and selling shares, the Japanese company still relies heavily on bank financing. Thus the American manager has to worry about his quarterly P & L and the effect current results will have on the company's stock. That is his report card. The Japanese manager, once he has satisfied his board and his banks about his long-range business plans, is relatively free to work them out, without the need either to push for short-term results or to constantly explain his situation to his board, not to mention the friendly securities analyst next door. . . .
>
> The bank is the one overseer that the company directors have to watch and heed. And here too comes the question

21. Kanji Haitani, *op. cit.*, 80–81.

of building up confidence. First, money will be loaned on a parent company guarantee or some other form of security. Gradually, assuming that the company does well—or shows promise—the local bank manager will feel more relaxed about his money. In Japan he is quite powerful. With good prospects, the company can secure loans from other banks, encouraging what can be healthy competition. As the banks scrutinize operations, get to know the people involved, and develop a feel for the business, they grow more helpful. If a long-range plan is explained to them and in their judgement makes sense, Japanese banks will wait a long time for the profits to come. . . .

Replacement is almost always internal, except in the case of near-disasters, when it is generally the firm's leading bank that steps in to save the situation (and its investment). This is what the Sumitomo Bank did with Toyo Kōgyō. It was the bank's responsibility to save the company, in contrast to the Chrysler case, where this charitable chore fell to the U.S. government.[22]

0.0 Sustained High Savings

Where do the banks get their funds to make the long-term loans to industry? From the savings of the Japanese people. Thanks to the bonuses paid two or three times a year, the working people get five or six months of salary at one time, a sizeable portion of which they put into savings:

In 1980, Japanese families saved 22 percent of their disposable personal income. The comparable American figure was 6 percent. For many years, Japan's gross domestic capital formation—the total of corporate, personal, and government savings—has approximated 40 percent, twice what it is in the United States and Europe. Actual private sector capital investment is about 20 percent of GNP, also twice what it is in the United States.[23]

Baranson points out an important incentive to investment: "The availability of investment capital to Japanese management is enhanced by the fact that interest is a pretax expense, whereas dividends are an after-tax drain. (In the early 1970s, Japanese firms were distributing to their shareholders an average 38 percent of

22. Gibney, *op. cit.*, 7, 79–80, 65–66.
23. *Ibid.*, 156.

after-tax earnings, as compared to 52 percent for U.S. corporations, according to the Boston Consulting Group."[24]

0.0 Reserves and Reinvestment

The lower dividends paid out to Japanese stockholders make it possible for the companies to retain much higher cash flow for expansion. The retention is in the form of reserves for several purposes—retirement funds, overseas investment losses, accelerated depreciation (on which taxes are deferred). Reserves and deferred dividend payments together generate a greater amount of cash flow for expansion.

The philosophy of the Japanese corporations favors ploughing back earnings into investment. This reinvestment is decentralized at Matsushita, which, as we pointed out, is the largest manufacturer of electric appliances in the world:

> Matsushita required that each autonomous division pay 60 percent of its pretax profits to the head office. . . . The remainder of a division's profit (about 40%) belonged to the division and was spent on such things as facility updating and expansion, production engineering and "self renewal", that is, bringing in a new generation of products to replace the older ones. . . . Of course, Matsushita has the ability to shift resources into new areas, using the 60 percent of income it receives from divisions. . . . Matsushita gives the divisions the major responsibility for doing that [reposition their products in tune with the marketplace].[25]

0.0 Investment in Research and Development

The administrative guidance that MITI exercises over growth industries usually involves government support of research and development in the chosen area. More than the actual amount expended, it is the fact of the governmental auspices in identifying and charting a long-term plan of growth for the particular industry that generates the R & D investment on the part of the industrial firms. In the 1950s, heavy industries received special attention and support. Investment in steel was encouraged by the MITI slogan, "Steel is the nation." Kenichi Ohmae, in *The Mind of the Strategist,* points out that MITI openly endorsed Kawasaki, which committed itself to a growth strategy and thereby stimulated the whole industry in an investment effort:

24. Jack Baranson, *The Japanese Challenge to U.S. Industry,* (Lexington, Mass.: Lexington Books, 1981), 11.
25. Richard T. Pascale and Anthony G. Athos, *The Art of Japanese Management* (New York: Warner Books, 1982), 56–57.

As a result, Japan as a nation has been able to produce high-quality steel at the lowest cost in the world. This competitive steel has been the underlying force behind Japan's current position in shipbuilding (50 percent of global tonnage construction), automobiles (30 percent of global units produced in 1980), and many other export-oriented industries, such as home appliances, machine tools, steel structure, and plants.

Today, MITI feels that the days of steel's dominance are about to end and is looking to very large scale integrated (VLSI) circuits to power the next industrial era. It has begun referring to VLSI as the "rice of industry," meaning that it feeds into all industries, much as rice is the basic daily food of all Japanese. . . . Many foreigners, critical of the Japanese government's "subsidies" to VLSI, mutter about Japan, Inc., or unfair competition. The truth is that the government chipped in a mere $130 million of the $320 million required by a four-year-long project. . . .

Apart from the rare investments of this kind, MITI has confined itself to openly endorsing vital R & D programs, mainly to develop alternative technologies for VLSI production. With this objective, five companies—Hitachi, Fujitsu, Mitsubishi, NEC, and Toshiba—formed an ad hoc cooperative in 1976, which dissolved itself in 1980 after four years of joint R & D efforts. . . . None of the five can afford not to take advantage of the fruits of those four years of joint effort.[26]

0.0 The Visible Hand

The administrative guidance, endorsement of Research and Development programs and of companies undertaking desired growth strategies, and the actual Research and Development investment itself are but expressions of the ethos of what may be termed the Visible Hand of MITI and the Japanese bureaucracy looking after its own. Chalmers Johnson's identification, in *MITI and the Japanese Miracle,* of the factors favoring Japan's industrial growth may be arranged as follows:

- Foreign currency quotas and controlled trade suppress potential domestic demand to the level of the supply capacity of an infant domestic industry.
- High tariffs suppress price competitiveness of a foreign industry to the level of a domestic industry.

26. Kenichi Ohmae, *The Mind of the Strategist* (New York: McGraw-Hill, 1982), 231–32.

- Large purchasing power of consumers is raised through targeted tax measures and consumer-credit schemes, thereby allowing them to buy the products of new industries.

- An industry borrows capital in excess of its borrowing capacity from governmental and government-guaranteed banks in order to expand production and bring down unit costs.

- Efficiency is raised through the accelerated depreciation of specified new machinery investment.

- Tax incentives for exports function to enlarge external markets at the point of domestic sales saturation.[27]

0.0 The Inventional/Innovational Factors of Japanese Success

From the brief survey of institutional and investmental factors, we turn to the factors of invention and innovation.

0.0 "Gain the whole world market first"[28]

In chapter 1, we saw that during the Industrial Revolution, territory stimulated technology. Colonial and other overseas trade gave British industry an unrivaled foreign market. The market (territory) stimulated the technology of the Industrial Revolution. We pointed out that, in the coming Robotics Revolution, the technology could well be providing the stimulus to find the territory, reversing the thrust of the Industrial Revolution. We pointed out that, in the coming Robotics Revolution (RR), the technology could well be providing the stimulus to find the territory, reversing the thrust of the Industrial Revolution.

As the robotics technology began emerging and appearing viable, Japan started developing her territory, her market. This practice is almost ingrained in the Japanese, who have to depend on trade for food and fuel. Gibney traces the preoccupation with the market to the post–World War II days:

> Because Japanese long-range planning in World War II had been such a failure, they now concentrated on planning and strategies, almost to a fault. Everything had to be thought out well in advance, tested—and above all—adapted to the needs of the marketplace, wherever

27. Chalmers Johnson, *MITI and the Japanese Miracle: The Growth of Industrial Policy, 1925 –1975*, (Stanford: Stanford University Press, 1982).
28. Gibney, *op. cit.*, 151.

that marketplace was. Thus the Japanese businessman appeared as the first world marketer. . . .

By contrast, [with the American manager], the Japanese manager would prefer to *gain the whole world market first* and then work on the salvation of his profitability through improvements in productivity. For what, to his way of thinking, does it profit a man to have a broader *pro forma* profit on sales in a small and dwindling share of the market? . . .

Although Japan's total exports rarely exceed 12 percent of the GNP—less than half the percentage of big exporters like West Germany, for example—they are an important 12 percent. . . . The fact that Japan must export to feed itself, literally, is *etched on the consciousness of every Japanese,* conspicuously including customs officials and bureaucrats in the economic ministries. . . .

The whole impression of Japanese exports and, now, investments resembles that of an army on the march. The huge multinational trading companies move ahead first, testing markets, arranging local contacts, buying up available supplies of raw materials and component parts, like mobile *reconnaissance* units. The overseas offices of the semigovernment JETRO [Japan External Trade Organization] assist in the work of *intelligence* and act as a backup for Japanese businesses when the main body comes to sell. Meanwhile the *government in Tokyo* lays down heavy artillery barrages in the form of tax credits, depreciation allowances, and R & D assistance as artillery support.[29] (emphasis added)

To gain the whole world market, the Japanese pursue a carefully-laid-out strategy. We will discuss the thoughts of one man, who is called by his native Japan, "Mr. Strategy."

0.0 Four Routes to Strategic Advantage

Ohmae describes a good business strategy as "one by which a company can gain significant ground on its competitors at an acceptable cost to itself."[30] This may be accomplished in four ways:

29. Gibney, *op. cit.* 49, 151, 164, 167–80.
30. Ohmae, *op. cit.*, 36.

Robotics/Artificial Intelligence/Productivity

1. Concentrate resources in a particular area (such as servicing or purchasing) to gain the most strategic advantage over competition
2. Build on relative superiority (that is, attract small users who would enlarge their use of copiers) by offering low-admission that competition cannot duplicate
3. Upset the status quo by rethinking common-sense notions (such as questioning why inventory should be stockpiled at the automobile factory)
4. Serve the user better in areas where competition will be slow to follow (for example, select film, optical systems, and mechanical systems as the three areas with the greatest potential for improvements).

Clients of McKinsey & Co. in Japan successfully pursue this strategy. Ohmae draws upon many real-life examples to elucidate his concepts. The key is that one should push to the fore the competitive advantage in one or two selected areas of customer concern and satisfaction with such a concentration of resources as to make it yield "functional competence" in specific segments.

> The secret of many Japanese corporations' success is their skill in sequencing the improvement of functional competence. In the 1950s and early 1960s many of them made heavy investments of both money and talented people in manufacturing engineering. Their production/technology, together with the advantage in labor cost they enjoyed at that time, constituted their principal source of strength. At this stage, their investments in research and development and overseas marketing were minor; for these they relied on imported technology and trading firms, respectively. Later they shifted their emphasis to quality control and product design capabilities. Today they are very active in basic research and direct marketing. At each phase they have been able to generate the money to reinvest in improving the next generation of functional strengths. . . .
>
> Recognizing its competitors' inability to introduce new products rapidly, Casio has adopted *a strategy of accelerating and shortening product life cycles*. No sooner had its 2 mm-thick, card-size calculator been introduced than Casio started rapidly bringing down the price, thus discouraging its competitors from following with a similar product. Within a few months, Casio introduced another model, which emits musical notes as the numerical keys are touched.

64

> In Casio's case, the functional strategy is to integrate de-
> sign and development into marketing so that consumers'
> desires are analyzed by those closest to the market and
> quickly converted into engineering blueprints. . . . Its
> competitors, all organized vertically on the assumption of
> a one- or two-year life cycle for this type of product, are
> at a severe disadvantage.[31] (emphasis added)

To gain the world market, the Japanese laid out and pursued flexible functional competence. Canon, Ricoh, Panasonic, and Pentax all wanted to acquire
global markets, but they were willing to forego their pride as original equipment
manufacturers (OEMs) and supply components or even whole products to be sold
under western trademarks. Had they not done so, they could not have competed
on many fronts with the western competitors. Once they achieved functional
competence in product quality and cost competitiveness, they would move on to
dispensing with OEM relationships. As Ohmae sees it, this sequencing—OEM
selling not under their own but under foreign trademarks, developing marketing
efficiencies, and then selling under one's own brandname—is Japanese marketing
wisdom.

0.0 Sell Better for Less
While Ohmae suggests four routes to strategic advantage, Matsushita has become
the largest electrical appliance manufacturer in the world essentially by concentrating on reaching functional competence in one area—production engineering. Converting the cost advantage realized from production engineering into lower prices
for the consumer, Matsushita captures the market and expands it. The philosophy
is to gain the whole world market first. Going against the 1930s' practice of lowering costs and charging higher prices, Matsushita developed an aggressive strategy to capture the wider market, outdoing Henry Ford:

> Matsushita, inspired by Henry Ford's pricing strategy
> for the Model T, grasped the concept of *aggressively
> pursuing market share,* gaining economies through manu-
> facturing experience, and *lowering price, thus establish-
> ing barriers to entry for competitors* who found the small
> margins unattractive. . . .
> At the heart of Matsushita's followership strategy is pro-
> duction engineering. . . . To this day, Matsushita rarely
> *originates* a product, but always succeeds in manufactur-
> ing it for less and marketing it best. . . . Their [Matsu-

31. *Ibid.,* 38.

shita's] 23 production research laboratories concept of 'research and development' is to analyze competing products and figure out how to do better . . .

Matsushita's underlying strategic assumption is that *profits are linked to growth* and that investments which promote growth will eventually pay off in profits *over the long term.*[32] (emphasis added)

0.0 Detailed, Pragmatic Planning

To convert the strategy of acquiring and expanding global markets, Matsushita insists on meticulous planning. Each division manager presents three plans every six months. The five-year plan deals with division responses to changes in its marketing and technological environment. The two-year plan discusses how the five-year plan demands will be met by plant capacity and new products. The most carefully scrutinized of the three plans is the third one, the "Program for the Next Six-Month Operating Period." Former IBM executive Lee Shevel, now manager of Matsushita Motorola, says:

> The planning system is beautiful — at least on a par with IBM's. It is very detailed and pragmatic. When a manager's six-month operating plan asserts he is going to increase capacity by x units and sales by y dollars per year, they nail it down every step of the way — when he is going to hire salesmen, who needs retaining, what production equipment goes on order, and how long it takes to debug the assembly line. There are few bald assertions; it's very, very carefully done.[33]

0.0 Profit Center Review of Plan Variances

The most detailed plan would come to nought, however, if deviations from it were not detected and corrected in time, or if the plan were not revised in light of new factors. At Matsushita, the controllers are each assigned to one of the over 100 operating divisions. Their role is not to act as a spy from headquarters, but, as a controller himself put it: "[to] act more like a wife in the traditional Japanese household. Like the wife, we work largely outside the public view but keep tabs on the finances and remind the division head how things are doing."[34]

Matsushita and Takahashi, his finance expert, together pioneered the world's first profit center, which is the division. The divisions forward monthly and quarterly data to the executive vice president for finance with explanations for each

---

32. Ohmae, *op. cit.*, 112–13, 117.
33. Pascale and Athos, *op. cit.*, 40–42.
34. *Ibid.*, 51.

variance from the six-month plan. Next, the division manager and staff spend several days at headquarters going over each item to prepare for senior management review. Senior management will review the plan itself in light of experience with it. An interesting source of motivation for the divisions to become better performers is the review of summary operating results in front of their peers. Their performance is ranked, from outstanding (A) to the poorest (D). The fact that one is in a category other than A acts as a powerful motivator, particularly when one is in C or D.

0.0 Common Thread of Productivity Triad

We have discussed what appear to be three key groups of factors in Japanese productivity, which are institutional, investmental, and inventional/innovational.

It is easy to take the position that east is east and west is west and never the twain shall meet. Keats could afford to leave it at that. We cannot. It would be absurd, though, to borrow the symbols of Japanese entreprise elements without their substance. It would be as absurd as donning a Japanese *obi,* worn with a kimono as a belt over an American tuxedo. We must ask what underlying factors make their productivity triad work and investigate the possibility of emulating and adapting them to enhance American productivity.

The hallmark of Japanese life is harmony which makes possible the coexistence of opposing forces. A new concept in Game Theory explicitly treats inherent oppositions in individual, group, national, and international life. Using that concept, we can illuminate the opposing interests involved when advanced, developing countries import high-technology, as symbolized by the robots, in order to export high technology products. The concept itself is developed next for its application to national players in Japan and in the U.S. in the realm of productivity and robotics.

PART II
Improving Productivity Through Integrated Manufacturing

4

Robots, CAD/CAM, and Other Automation Technologies in Integrated Manufacturing

OVERVIEW: Japan and the United States differ on the very definition of industrial robots. The Robot Institute of America emphasizes reprogrammability and multiple task performance, while the Japan Industrial Robot Association (JIRA) emphasizes intelligent functions, meaning at least one of the following: judgement, recognition, adaptation, or learning. Irrespective of the definition, Japan is the front-runner in the industrial robot population: JIRA figures credit Japan with 62.6 percent and the U.S. with 18.1. Some seventy second-generation Direct Numerical Control (DNC) machines are in operation in Japan, in which the program of instructions is transmitted from the large central computer directly to the machine tools.

Is robotics really necessary for American manufacturing? Yes, according to the National Academy of Sciences. Their Committee on Computer-Aided Manufacturing points out that manufacturing is responsible for two-thirds of wealth-creation. By reducing the cost of manufacture, computer-integrated manufacturing offers the greatest potential to reduce the cost of generating wealth. Chapter 4 discusses recent developments in four potential areas of savings in manufacturing: *assembly*, *feeding*, *programmable assembly*, and *design*.

Actual cost savings are discussed for the United States, Japan, and West Germany. *The American Experience*: John Deere Tractor Works won the first Annual Award of the Society of Manufacturing Engineers for exemplifying CAD/CAM (Computer-Aided Design/Computer-Aided Manufacturing). They report reducing by 25 percent the work-in-process inventory, and an additional $6 million by employing Group Technology (GT), in which parts to be machined are grouped into families according to common characteristics. *The Japanese Experience*: number 2 automaker Nissan's Zama auto plant uses Flexible Line Assembly to customize cars, e.g., incorporating widescreen wipers in the headlights of cars sold in Denmark. In Nissan's new plant in Smyrna, Tenn., the fully integrated

monitoring and control system is estimated to save 40 percent and more. *The West German Experience*: Messerschmidt reports savings by using FMS (Flexible Manufacturing System): 44 percent reduction in the number of machines, 30 percent reduction in the work force, 25 percent increase in throughput, and 24 percent increase in annual returns. Controls being at the very heart of robotics, chapter 10 treats them at great length.

0.0 The Robots Are Coming! The Robots Are Coming!

The unprepared meeting the unexpected is the theme of the 1960s American movie, "The Russians are coming! The Russians are coming!" We see the Russian commander of the submarine that accidentally surfaced near a sleepy coastal town in the northeastern United States posturing uncertainly, fueling the fires of fear of imminent attack in the bewildered population, which rushes to raise the barricades. Their mutual fear is heightened by a total ignorance of each other's language.

Today, the robot can be perceived as a lumbering enemy bent on destruction of one's dearly-held job and, indeed one's very way of life. The Luddites in England in 1779 so viewed the modern engines of the day and went about vigorously destroying them. It took parliamentary laws penalizing such destruction by death to stave off total annihilation. More important than the laws themselves was the expanding trade in overseas territories, which restored and significantly increased employment in England a decade later.

Instead of viewing the robot with alarm, it could be considered properly a modern slave, to be employed widely in enterprises ranging from automotives to cosmetics, releasing men and women for higher pursuits. Whether the robot is perceived as a menacing enemy or a willing slave depends on understanding the language of robotics. If we do not understand, Bernard Shaw's observation will apply, that America and England are two countries separated by a common language. Today and tomorrow will be the best of times and the worst of times, separated by the common language of high technology.

0.0 Will the Real Robot Please Stand Up?

Turning to robots as an embodiment of high technology, we find the language to be confusing. There is no internationally accepted definition of *industrial robot*. We need, therefore, to consider different definitions and to bear them in mind when making comparisons.

The Robot Institute of America emphasizes reprogrammability, and variety of tasks in its definition: "A reprogrammable, multifunctional manipulator de-

signed to move materials, parts, tools or specialized devices through variable pro-grammed motions for the performance of a variety of tasks."[1]

On the other hand, the Japan Industrial Robot Association (JIRA) empha-sizes intelligent functions: "A mechanical system which has flexible motion func-tions analogous to the motion of living organisms or combines such notions with intelligent functions, and which acts in response to the human will. In this con-text. intelligent functions mean ability to perform at least one of the following: judgement, recognition, adaptation, or learning."[2]

If one of the four—judgment, recognition, adaptation, or learning—is indis-pensable in qualifying a machine as a robot, Japan's own count of industrial ro-bots appears to be mildly exaggerated. Based on surveys of robots in different countries (table 4.1), each one of which uses its own definition, JIRA credits Ja-pan with 14,150 IR (62.6%), the U.S. with 4,100 (18.1%), West Germany with 1,420 (6.3%), and Sweden with 940 (4.1%).

The issue of judgment, recognition, adaptation, or learning is no semantic squabble; it is at the heart of robotics itself. Even the staunchest advocate of ro-bots would hardly suggest that what humans consider to be judgment is now vested in robots. There is some question as to whether such judgment is even replicable. Taking the literal sense of the term, we see that western jurisprudence requires a court, the seat of judgment, to have three elements: equity, good con-science, and law. When we acclaim Solomon's judgment, it is his own particular mixture of the three elements that we acclaim.

If it is not devoutly to be desired that the robot should be "sober as a judge," so that a contemporary Shylock should exclaim, "A Daniel has come to judg-ment!", we need to specify what indeed it is that we look for in a robot, now and in the future.

0.0 Change Symbols, Instead of Machines, to Manufacture: Numerical Controls, Instead of Re-Tooling

Henry Ford accomplished mass production through fixed automation, each ma-chine repeating the same job. When the job changed, either the machine had to be retooled or a different machine had to be used.

Today, robots offer versatility, permitting each machine to function as many machines. By changing the instructions, the same robot could be d)ing almost such divergent functions as digging ditches one moment and sewing lace curtains the next. (None is performing such feats at the moment, though.)

Robots do perform different tasks, however, as Grossman points out:

> I should mention that one location in IBM is actually us-ing a robot for small batch assembly. Every few hours,

1. Robot Institute of America, Dearborn, Michigan.
2. Japan Industrial Robot Association, Tokyo, Japan.

Table 4.1
Industrial Robots Population by Countries
(Excluding Manual-Manipulator and Fixed-Sequence Robot)

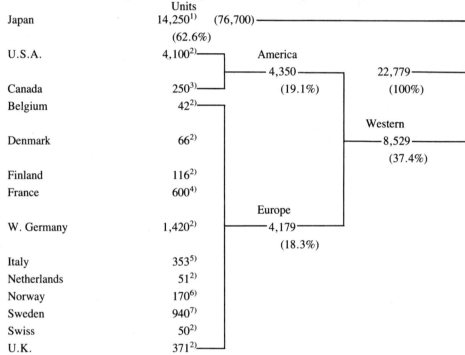

	Units			
Japan	14,250[1] (76,700)			
	(62.6%)			
U.S.A.	4,100[2]	America	22,779	
		4,350	(100%)	
Canada	250[3]	(19.1%)		
Belgium	42[2]			
			Western	
Denmark	66[2]		8,529	
			(37.4%)	
Finland	116[2]			
France	600[4]			
		Europe		
W. Germany	1,420[2]	4,179		
		(18.3%)		
Italy	353[5]			
Netherlands	51[2]			
Norway	170[6]			
Sweden	940[7]			
Swiss	50[2]			
U.K.	371[2]			

Remarks: 1. JIRA Survey (1980–end), ( ) including Manual-Manipulator and Fixed-Sequence Robot
2. RIA Survey (1981)
3. National Research Council (1981–end)
4. Jetro, Paris Survey (1981, 11)
5. Italian Industrial Robot Association (1981)
6. RIA Survey (1980)
7. Swedish Computers and Electronics Commission, Ministry of Industry (1981)
British Robot Association Survey (1981 Dec.);

U.S.A.	5,000
Germany	2,300
Sweden	1,700
U.K.	713
France	600
Italy	450

Source: Japan Industrial Robot Association, Tokoy, Japan, May 1982.

the robot is provided with a different workstation, a new
program is loaded, and the robot changes to a new task.[3]

3. David G. Grossman, "A Decade of Automation Research at IBM" (Yorktown Heights, New York: IBM Resarch Center, RC 9062, #39592, 1981), 6.

Several essential developments paved the way from fixed automation to supervisory control.

The jet aircraft of the early fifties motivated the U.S. Air Force to help invent a machine to do the extensive milling required to build such aircraft. In March 1952, MIT laboratories demonstrated the first numerical control (NC) machine. With NC, the step-by-step instructions are most often written as numbers, which are read by the controller unit, which, in turn, moves the machine tool. The program could be written in the form of holes on a punched tape, the rate at which the tape is fed determining the speed of the machining operation.

The punched tape, being subject to wear and tear, becomes inaccurate in operation. This led to searches for tapes of mylar and aluminium foil, which were more reliable but more expensive. The tape and the tape reader are omitted in direct numerical control (DNC), in which the program of instructions is transmitted from the computer memory directly to the machine tool. The DNC appeared in the late sixties.

0.0 Second-Generation Direct Numerical Control in Use in Japan

While DNC is run by a large central computer, computer numerical control (CNC) typically uses a minicomputer as the central unit. DNC runs a large number of machines, CNC runs only one. DNC controls hundreds of machines far away, CNC controls only physically proximate machines.

The main advantage of NC, DNC, and CNC is that they reduce nonproductive time in manufacturing, which are principally tool changes and job setups. They do not *compensate*, however, for variations in machine operations. This is the contribution of *adaptive control* (AC), invented in 1962 also through the support of the U.S. Air Force. AC measures one or more process variables, e.g., temperature, cutting force, etc., and manipulates feed and/or speed.

The introduction of AC into NC paved the way for the industrial robot, which handles materials (*materials handling*), instead of *metal processing* as with NC. Robots were born in the United States and raised elsewhere. By 1967, both the United Kingdom and Japan imported Unimate, made by Unimation, Inc., and Versatran, made by AMF, Inc.

Some seventy DNCs are in operation in Japan today. The first DNC system was installed in 1968 at a National Railway factory that repaired vehicle parts. The central computer prepares the schedule for machining by seven NC lathes. Transferring of parts, however, and loading and unloading parts have to be done manually. The second-generation NC, installed in 1972, is equipped with transferring, loading, and unloading capabilities as well. When one of the lathes reaches the final stage of a machining process, the robot approaches the lathe and waits. When the machining is finished, the robot unloads the piece and places it on the work feeder. Professor Hiroyuki Yoshikawa of Tokyo University, who has been a

member since 1977 of the Steering Committee of the national project, "Flexible Manufacturing System Complex with Laser," describes the robot:

> The robot's hand plays a most important role in improving flexibility. The robot has nine degrees of freedom in motion, with an accuracy of $\pm$ 1 millimeter. The hand has three fingers that act as chuck keys on workpieces of different shapes. It rotates the workpiece without disturbing other equipment. The robot runs at 45 meters/minute, and the rotations are at about 50 degrees/second. The robot is driven by hydraulic motors with encoders, and the hand is driven by step motors. Driving signals are usually produced in the computer by a teaching system. A worker manually operates the robot through a robot-operating panel, and the motion is stored in the computer. Afterwards, the system can call the motion data at any time by assigning the name of the data taught and stored.[4]

Notice that the second-generation DNC has nine degrees of freedom. This is three more than the six basic motions, or degrees of freedom, of a robot—three arm and body motions and three wrist movements. In Taiwan, Teco Electric & Machinery Co. is manufacturing the ITRI-E(lbow) robot developed by the government-funded Industrial Technology Research Institute (ITRI). The ITRI robot has four degrees of freedom, a loading capacity of six kilograms, and is suitable for moving, lifting, classifying, and assembling operations in electronic plants, garment factories, plastic processing, and machinery industries.

0.0 The Economic Imperative of Robotics for Integrated Manufacturing

Is robotics really necessary for American manufacturing?

Chairman and chief executive officer of the American Motors Corporation, W.P. Tippett, presented some disturbing facts:

> In 1962, we imported a total of $7.6 billion worth of manufactured goods. Last year [1981] the total was $143 billion, an increase of almost 2,000 percent over two decades. . . .
>
> In the past two-and-a-half years, employment in the U.S.

---

4. Hiroyuki Yoshikawa, "State of Art in Japan," in *Computer-Aided Manufacturing: An International Comparison*, Committee on Computer-Aided Manufacturing, Manufacturing Studies Board (Washington, D.C.: National Academy of Sciences, 1981), 10.

auto industry has declined by 34 percent: in the steel industry, it has declined 48 percent. Yet during the same period, imports of automobiles rose 43 percent, and imports of steel increased 46 percent. . . .

The service sector of our economy now accounts for two-thirds of total GNP and three-fourths of the total work force!

I submit that *no country can hope for long-term economic growth if two-thirds of its GNP is accounted for by service industries*. When you sell an insurance policy or a hamburger, you simply don't create the ripple of growth that happens when you manufacture an automobile, a steel pipe, or even a toothpick. The Japanese have recognized this since the early 1930s, when they began taking steps to force foreign companies out of Japan in order to create their own auto industry and strengthen their other basic industries.

Even more important, you can't hope to remain a world political leader or even guarantee your national defense if you rely on other countries for the bulk of your manufactured products. How long can the United States remain the leader of NATO and the Pacific alliance if we continue to allow our manufacturing base to dwindle? What becomes of our defense capability if we have to rely on foreign countries for weapons, vehicles and parts?[5] (emphasis added)

0.0 Manufacturing's Declining Share in GNP

Tippett's statistics on the dwindling share of manufacturing of GNP appear to be more optimistic than those of the Department of Commerce. We see from table 4.2 that in 1947, manufacturing of durable and nondurable goods together accounted for $60 billion, which represented 30 percent of the national income. In 1980, the percentage declined to 24 percent, the remaining 76 percent being accounted for by nonmanufacturing industries. Tippett's point is that the manufacturing base is the source of income, and as such, constitutes income-producing capability — *wealth*.

0.0 Manufacturing's Significant Share in Wealth-Creation

Considered as wealth-creating, manufacturing is critical to the economy. The Na-

---

5. W. P. Tippett, "Getting Our Basic Industries Back on Track," *Vital Speeches of the Day*, Speech delivered 16 December 1982, 263, 264.

Table 4.2

National Income Originating in Manufacturing: 1947 to 1980

(In billions of dollars, except percent. Prior to 1963. excludes Alaska and Hawaii. Refers to national income without capital consumption adjustment. Data represent net value added at factor costs. See *Historical Statistics, Colonial times to 1970*, series F 230, for single-year figures prior to revisions issued in 1981)

Item	1947	1954	1958	1963	1967	1972	1974	1975	1976	1977	1978	1979	1980
Total manufacturing	60	95	108	144	194	252	301	318	367	411	404	515	527
Percent of national income	30	31	29	29	29	26	26	25	26	26	26	26	24
Durable goods	32	55	62	86	119	154	181	186	217	249	286	315	312
Percent of total	53	58	58	60	61	61	60	58	59	60	62	61	59
Nondurable goods	28	40	46	57	75	98	121	132	150	163	178	199	216

Source: U.S. Bureau of Economic Analysis, *The National Income and Product Accounts of the United States, 1929-76,* and *1976-79,* and *Survey of Current Business,* July 1981.

tional Academy of Sciences would agree. A Report by the Committee on Computer-Aided Manufacturing of the Academy says:

> Manufacturing is currently responsible for about two-thirds of the wealth created annually in the United States.[6] Creation of wealth is essential to the advancement of the standard of living, quality of life, and level of employment. These factors may be increased or improved through the reduction of the cost of creating wealth. Thus, reducing the cost of manufacturing is clearly of major importance to the economic and social well-being of this country.
>
> Technological advances provide the greatest contribution to manufacturing productivity increases and therefore to reducing the cost of generating wealth.[7] *Computer-integrated manufacturing*, still at a very early stage of its development and implementation, has already shown clearly that it *has far greater potential* to reduce manufacturing costs, and reduce them dramatically, *than any other known technology*. It follows, then, that public funding in this country ought to be directed toward research, development, and implementation of computer-integrated manufacturing.[8] (emphasis added)

0.0 A Question of National Defense

Tippett provides strong motivation for improving the manufacturing capability of the United States when he asks "What becomes of our defense capability if we have to rely on foreign countries for weapons, vehicles, and parts?" We noted earlier that the jet aircraft of the early fifties provided the motivation to the U.S. Air Force to help invent a machine to do the extensive milling required to build such aircraft. The support of the Air Force was also responsible for the invention in 1962 of Adaptive Control, which compensates for the variation of machine operations from the norm.

The first numerical control machine was demonstrated at MIT in March, 1952. In 1969, computer numerical control (CNC) stored-program in a minicomputer replaced the hard-wire control unit in NC. DNC also appeared in 1969. In

6. Society of Manufacturing Engineers, "The National Role and Importance of Manufacturing Engineering and Advanced Manufacturing Technology," Dearborn, Mich., May 8, '78.

7. David L. Morrison and Keith E. McKee, "Technology—for Improved Productivity," in *Manufacturing Productivity Frontiers*, Vol. 2, No. 8, June, 1978, 1. The authors average the contribution of technology to productivity growth estimated by John Kendrick (72%), Edward Denison (62%), and Laurits Christensen, Dianne Cummings, and Dale Jorgenson (44%) to arrive at 59%.

8. The Committee on Computer-Aided Manufacturing, Manufacturing Studies Board, *Report on Activities, January 1980 to June 1981*, (Washington, D.C.: National Academy of Sciences, 1981), 5.

DNC, a central computer transmits the program of instructions directly to hundreds of distantly located machine tools. DNC is CAM (Computer-aided Manufacturing). Second-generation CAM in the form of flexible manufacturing systems (FMS) made their appearance in the early seventies. They were distinguished by their capability for automated materials handling.

The Air Force felt that CAM could benefit from an accelerated effort. The Air Force System Command's Materials Laboratory at Wright-Patterson Air Force Base formally started the Integrated Computer-aided Manufacturing Program (ICAM) in April 1976.

In the review of ICAM after five years, the Committee on CAM observed that the ICAM program is providing seed money for the introduction of information systems and production equipment such as robots, manufacturing cells, automated forming equipment, and material management and control software. They recommend the following priority to the nine thrust areas:

1000	Manufacturing and Architecture	
2000	Fabrication	Priority 1
3000	Data Base and Data Automation	
4000	Design/Manufacturing Interaction	Priority 2
5000	Planning and Group Technology	Priority 3
6000	Manufacturing Control & External Interfaces	
7000	Advanced Fabrication and Assembly	Priority 4
8000	Simulation, Modeling and Operations Research	
9000	Materials Handling and Storage	Priority 5
0000	Test, Inspection, and Evaluation	Priority 6

The Committee recognized the economic imperative of integrated manufacturing.

A new, highly dynamic period of CAM has begun in all industrialized countries of the world. By the turn of this century, the extent to which CAM principles are applied will be the determining factor of the status of a nation's industries. . . .

The path to this factor is dual: partly bottom-up, partly top-down. The bottom-up approach is through the construction of increasingly sophisticated machining cells which may be flexibly linked to one another by mainly software techniques. These cells, which allow unmanned night-shift operations, are in themselves competitive products. The top-down approach requires the thoughtful and

bold development of the overall factory of the future and
its testing through major experimental projects.[9]

0.0 Comprehensive Economic Justification for ICAM

The reference to an "overall factory of the future" underscores the fact that robotics-based manufacturing can hardly be justified strictly on the basis of manufacturing economies. The successive stages of coverage in economic evaluations of applications of CAM recommended by the Committee on CAM, reproduced in figure 4.1, start with *materials handling inventory*, shown as subsector B. Next come the *operations immediately preceding and following materials handling*, designated as Operations A and C. These operations impact on the other planning, control, and service functions primarily related to manufacturing, sketched in the upper part of the figure. The purpose of manufacture is sale; therefore, the double-arrow between manufacturing and sales is most germane to the realistic discussion of the "overall factory of the future." The counterpart of sales is purchasing, which enables manufacturing operations to be carried out. The manufacturing-purchasing link is also double-arrowed.

0.0 Potential Area of Savings in Manufacturing Costs: Assembly

How much room is there for savings in manufacturing costs?

Nevins and Whitney of Charles Stark Draper Laboratories present the data reproduced as table 4.3 showing the distribution of production costs. In Industry 3522: Farm Machinery and Equipment, 30.23 percent of direct labor is involved in production, but only in 7.07 percent of the value of shipment. Similarly, in Industry 3717: Motor Vehicles and Parts, 33.45 percent of direct labor used for assembly represents only 4.75 percent of the value of shipments:

> If we were to completely automate the assembly functions in the two industries we would only change our total costs by 7 percent for farm machinery and 4.7 percent for the auto industry. This, of course, assumes we know how to do so. [Table 4.4] illustrates this perhaps more dramatically. Here, not only are production functions changed, but also non-production functions—"white collar workers"—like engineering. Again, dramatic changes in the specific function cause only *minor—a few percent— modification in the final sum. . . .*
>
> Rather, what is needed is a systematic multi-discipline approach if one is going to achieve the kind of changes

---

9. Committee on Computer-Aided Manufacturing, *op. cit.*, 3, 41.

Figure 4.1
Successive Stages of Coverage in Economic Evaluations of Applications of
Computer-Aided Manufacturing

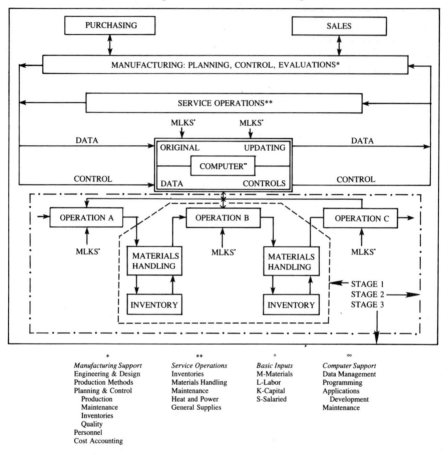

*	**	°	∞
Manufacturing Support	*Service Operations*	*Basic Inputs*	*Computer Support*
Engineering & Design	Inventories	M-Materials	Data Management
Production Methods	Materials Handling	L-Labor	Programming
Planning & Control	Maintenance	K-Capital	Applications
Production	Heat and Power	S-Salaried	Development
Maintenance	General Supplies		Maintenance
Inventories			
Quality			
Personnel			
Cost Accounting			

Source: Committee on Computer-Aided Manufacturing, Manufacturing Studies Board, *Report on Activities January 1980 to June 1981* (Washington, D.C.: National Academy of Sciences, 1981), 35.

needed to cause a significant effect on industrial productivity.[10]

0.0 Potential Area of Savings in Manufacturing Costs: Feeding
If a part can be automatically fed and oriented, it can usually be automatically as-

10. J. L. Nevins and D. E. Whitney, "Assembly Research," in *Automatica*, Vol. 16, (London: Pergamon, 1980), 596.

sembled. The UMass system, therefore, considers the difficulty of feeding and orienting the parts of an assembly. In 1972, Geoffrey Boothroyd, professor of mechanical engineering at the University of Massachusetts began to develop a classification and coding system for parts feeding for assembly, when one of his former graduate students, Alan Redford, of the University of Salford, undertook similar research across the Atlantic:

> Redford points out that it is not so much the cost of the robot as the cost of the feeding system which makes robotic assembly uncompetitive compared with a special-purpose assembly machine at one end of the scale and assisted manual assembly at the other end. . . .
>
> Any automatic feeding system is five to 10 times more costly than it would be if associated with a special purpose assembly machine, simply because a typical cycle time for an assembly machine is 2–3 seconds, whereas a multi-arm robot carrying out serial assembly of one part after another is likely to have a cycle time in the region of 20–30 seconds. . . .
>
> Redford believes that really economical batch production using serial manufacture with multi-arm robots is still a few years away. For the present and immediate future he sees the best opportunity for automated manufacture in what he calls programmable assembly—using versatile pick and place units in a line.[11]

0.0 Potential Area of Savings in Manufacturing Costs: Programmable Assembly

Programmable assembly, the aggregation of parts, is one of the two principal areas of assembly research, the other being part-mating, which occurs when two or more parts interact during assembly. The first experimental programmable assembly system was reported in 1973 by Kawasaki to assemble small gasoline engines, but it was subsequently abandoned.[12] The second was reported in 1975 by Olivetti, assembling small pieces with typewriter parts.[13] The third was reported

11. Jack Hollingum, "Automated Parts Feeder Design Cuts Cost of Robot Assembly," *Assembly Automation*, Nov. 1982, 215.

12. K. Seko and H. Toda, "Development and Application Report in the Arc Welding and Assembly Operation by the High-Performance Robot," *Proceedings of the Fourth International Symposium on Industrial Robots*, Nov. 1974, 19–21, 487–496.

13. A. d' Auria and M. Salmon, "'SIGMA' An Integrated General Purpose System for Automatic Manipulation," *Proceedings of the 5th International Symposium on Industrial Robots*, Sep. 1975, 185–202.

Table 4.3
Distribution of Production Costs

Category	SIC 3522 Farm Mach. & Eq.		SIC 3541 Machine Tools		SIC 3651 Radio & TV Sets		SIC 3717 Motor Veh. & Parts	
	% Labor	% Value	% Labor	% Value	% Labor	% Value	% Labor	% Value
Function costs	**100.0**	**23.39**	**100.0**	**37.51**	**100.0**	**15.92**	**100.0**	**14.19**
Non-production	42.68	9.98	32.28	12.11	52.67	8.38	39.40	5.59
Engineering	*5.63*	*1.32*	*7.17*	*2.69*	*20.10*	*3.20*	*5.45*	*0.74*
Production	55.98	13.10	66.49	24.94	45.94	7.26	59.07	8.38
Parts fabrication	*20.23*	*4.73*	*50.16*	*18.81*	*8.38*	*1.33*	*16.59*	*2.36*
Assembly	*30.23*	*7.07*	*10.98*	*4.10*	*23.90*	*3.80*	*33.45*	*4.75*
Inspection	*5.52*	*1.29*	*5.40*	*2.03*	*13.32*	*2.12*	*9.02*	*1.28*
Miscellaneous	4.56	0.31	1.23	0.46	1.73	0.28	1.53	0.22
Material costs		**54.05**		**35.35**		**61.54**		**64.81**
Quality costs		**1.5**		**1.5**		**1.5**		**1.5**
Other costs, profit, etc.		**21.06**		**25.64**		**21.05**		**19.50**

Design for efficient automatic assembly may require additional material. The balance between material and assembly costs determines how much the designer can afford (Nevins, Whitney, & Graves, SME, pub. MS82-125)

Table 4.4
Illustrative Impact Calculation Results

ILLUSTRATIVE EXPERIMENTS

Cost Elements	Farm Machinery and Equipment (SIC 3522)	Machine Tools, Metal Cutting Types (SIC 3541)	Radio and TV Receiving Sets (SIC 3651)	Motor Vehicles and Parts (SIC 3717)
Labor				
Staff				
Employee Relations				
Finance				
Marketing			+ 5.0	+ 5.0
Computer Operations				
Engineering	− 20.0	+ 20.0	+ 5.0	
Purchasing				− 10.0
Material Control	− 5.0			
Material Movement	− 2.0			
Fabrication				− 10.0
Assembly				− 10.0
Inspection			− 90.0	
Material		− 10.0		
Warrantee Cost			− 50.0	
Inventory	− 5.0	− 10.0		− 10.0
Fixed Assets			+ 0.1	
Change in Product Cost	− 3.0	− 3.0	− 2.5	0.7
Change in Asset Requirements	− 1.2	− 3.8	0.0	− 2.8

Note: All values are expressed as a percentage change in the ratio of costs to value of shipments.

in 1977 by Unimate, the Unimate-Ford U-6000, already in preproduction proto-type.[14]

The first programmable assembly experiment which generated valuable data on the system performance was reported by Nevins and Whitney in 1980. Nevins and co-workers developed a programmable test assembly bed in which the single electric arm for industrial assembly is supported by sensors, tooling, parts feed-ing, minicomputer systems and a specifically developed software system. The re-sults are shown in table 4.5.

> The economic model for programmable assembly systems identified a price-time relationship as an important tool for the selection or design of assembly robots to be used in programmable assembly systems. The price-time prod-uct is defined as the cost of the assembler (including in-stallation in the system) multiplied by the average single part assembly time for that assembler. Thus, the price-time product measures the effects of the cost of the as-sembler and its speed. . . .[15]

> Figure [4.2] illustrates the region of economic importance for programmable assembly based on . . . cost assump-tions listed in table [4.6]. Two points are to be noted. First, there is a general region for programmable assem-bly controlled by the yearly volume, the product of as-sembly station cost and assembly station time, and the payback period. Second, the boundary values among the various regions can be quantitatively examined by sensi-tivity analysis for specific applications both for economic and technological constraints and for indicating research directions and the priorities for carrying them out.[16]

These results were obtained on the basis of the economic-technological model that Lynch developed.[17] The results obtained, shown in table 4.5 are an initial approximation. To refine the measurements, Whitney and Graves developed a linear and integer programming model.[18]

Despite the progress made in identifying economically viable regions for the

14. M. J. Dunne, "An Assembly Experiment Using Programmable Robot Arms," *Proceedings of the 7th International Symposium on Industrial Robots*, Oct. 1977, 387–399.

15. J. L. Nevins and D. E. Whitney, *op. cit.*, 602–3.

16. J. L. Nevins and D. E. Whitney, "Research on Advanced Assembly Automation," *Computer*, Dec. 1977, 34.

17. P. M. Lynch, "Economic-Technological Modeling and Design Criteria for Programmable As-sembly Machines," Ph.D. Thesis, Dept. of Mech. Engg., MIT, June 1976.

18. D. E. Whitney and S. Graves, Seventh NSF Grantees' Conference on Production, Research and Technology, Cornell University, Ithaca, N.Y., 1979.

Table 4.5
Results from Experimental Programmable Test Bed

---

(1) Adaptable, programmable assembly of a real industrial product was accomplished:

adaptability was demonstrated by the system's ability to absorb position error of 1.25 mm (0.05 in.) or more due to part variations and uncertainty in the location of part feeders;

programmability was demonstrated by ability to program different assembly sequences using substitute tools in place of broken tools or new "tricks" to improve system performance;

approximately 75 alternators were assembled to test the system.

(2) Performance of the system was thoroughly documented

(3) Concept of engineered compliance was shown to be sufficient to perform a variety of difficult single-direction tasks that could either not have been performed at all, or at least with the necessary reliability.

(4) The software and teaching technique appear adequate for this class of assembly. For this problem planning was done off-line and explicit locations of tools and feeders was taught on-line, i.e. on the shop floor.

Table 4.6
Basic Cost Assumptions

Number of parts	= 10
Programmable station price	= $30,000
Tooling per part	= $7500
Transfer machine cost per part	= $30,000
Part station time	= 3 SECONDS
Payback period	= 2 YEARS TO 4 YEARS
Labor cost	= $7.50/HR TO $10.00/HR
Manual assembly station time	= 7 SECONDS
Number of seconds per year	= 1.152×10^7

---

introduction of assembly into manufacturing, Nevins feels that there is far to go. "You can't automate something you don't understand,"[19] he says, after a decade of directing the robot research center at the Charles Stark Draper Laboratory. He feels that robomation requires a much deeper understanding of the processes hitherto taken for granted, such as parts-mating and assembly. The design factors pe-

19. Doug McCormick, "Making Points with Robot Assembly," *Design Engineering*, July–August 1982, 23.

Robotics/Artificial Intelligence/Productivity

Figure 4.2

Economic Comparison of Manual, Fixed Automation and
Programmable Assembly Systems

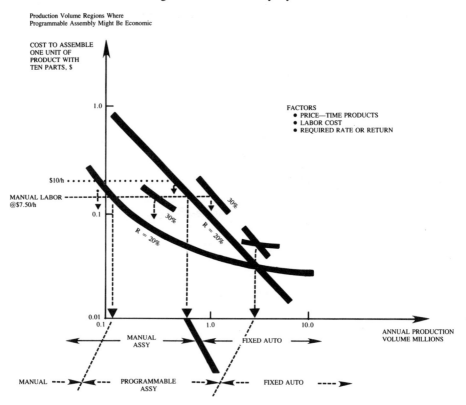

culiar to robomation have yet to be cofidied, and they have to be codified from
hundreds of pages of research reports.

The practitioners are even more skeptical. Peter Rogers, manager of product
development at Unimation, which manufactures Programmable Universal Manipulator for Assembly (PUMA) says: "No one in the world has any experience with
robot assembly . . . as a fully programmable *system* for producing an assembly or
subassembly."[20]

0.0 Potential Area of Savings in Manufacturing Costs: Design
In 1962, the British Institution of Mechanical Engineers was concerned that while
much work was being done on machine tools, little was being done on assembly.

20. *Ibid.*

Figure 4.3
Performance of Typical Subassembly and Final Assembly

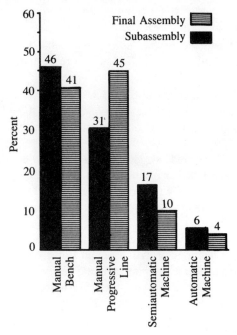

In response, Alan Redford, then a research student at the University of Salford, started work on vibratory bowls under the supervision of Geoffrey Boothroyd, now professor of mechanical engineering at the University of Massachusetts. In 1982, the University of Massachusetts published a widely-acclaimed volume, *Design for Assembly—A Designer's Handbook*. We reproduce Boothroyd's graph as figure 4.3. showing that only 6 percent of subassembly and 4 percent of assembly is automated:

> The first requirement when contemplating automatic assembly is to consider the product design and its suitability for assembly automation. . . . Two main factors affect the cost of assembly of a product: (1) the ease with which component parts are handled and assembled; and (2) whether the minimum number of parts has been used in the product. . . . In practice, it is found that the insertion process consumes the greatest cost in manual assembly as most small parts are quite easy to grasp and manipulate. In automatic assembly however it is parts handling that presents the greatest difficulty. . . .

To use the [University of Massachusetts] approach three factors are needed:

(a) a means for determining the *theoretical minimum number of parts* (NM);

(b) an estimate of the *minimum* possible handling and minimum *insertion time* (TH and TI respectively) for one part; and

(c) for each component part of the design under consideration, an estimate of the *extra time* (above the minimum) resulting from its particular design features.[21]

Boothroyd simplifies and categorizes basic geometries to apply the foregoing approach to determine whether or not automatic assembly is feasible. Design for manual handling and assembly is also discussed in the *Handbook* at length.

Boothroyd emphasizes the role of design:

There is no way robots can be economically applied unless you begin tailoring the design to the demands of assembly. Unfortunately, formulating design rules for robot assembly will be the last stage: you need to know what kinds of grippers, feeders, and so on you'll be using BEFORE you begin to evaluate the design.[22]

It will be some time, however, before the standards for grippers, feeders, etc. are set. Part of the reason is that there is little theory of assembly.[23] The literature deals mostly with specific details of assembly, such as methods of fastening, machines for orienting, designs of transfer mechanisms, etc.[24] Government-funded efforts in the U.S. support the study of hand-eye coordination. At the Charles Stark Draper Laboratory, the work on programmable assembly machines emphasizes good low-level control and tactile sensing.[25] Another aspect of sensory robotics, vision, is the emphasis at Stanford Research Institute.[26] The theory of assembly itself is studied at Stanford University.[27]

21. G. Boothroyd, "Design for Productibility—The Road to Higher Productivity," *Assembly Engineering*, Mar. 1982, 95, 93.

22. Doug McCormick, *op. cit.*, 25–26.

23. W. V. Tipping, *An Introduction to Mechanical Assembly*, (London: Business Books, Ltd., 1969).

24. Institute of Production Engineers, *Automated Assembly*, Vols. 1–18, London, 1970.

25. J.L. Nevins, D. E. Whitney, and S. N. Simonovic, "Report on Advanced Automation: System Architecture for Assembly Machines," Draper Lab Report R674, Cambridge, Mass., 1973.

26. C.A. Rossen, "The Future of Automation," *Naval Research Reviews*, Aug. 1973, 1–15.

27. R. Bolles and R. Paul, "The Use of Sensory Feedback in a Programmable Assembly System," AI Memo 220, Stanford University, Stanford, Cal., Oct. 1973.

0.0 Toward the Automated Factory: CAD/CAM Integration—the American Experience

The industrial robot is a reprogrammable materials handler. Manufacturing, however, is more than materials handling. The contribution of robotics will apply to the entire manufacturing operations, not merely to materials handling. We should, therefore, look at the current concepts of integrated manufacturing, of which robotics is an integral part.

John Deere Tractor Works of Deere & Co. was the first winner of the Annual Award given by the Society of Manufacturing Engineers for the project best exemplifying CAD/CAM integration. Five years in development, the project became fully operational in May 1981, *replacing by 25 percent the work-in-process inventory required.*

As an acknowledged success, Deere Tractor operations may be studied in terms of several elements of CAD, CAM, robotics, GT, and CIM which contribute to its success.

0.0 Distributed Hybrid Network

The 2,200 Deere dealers in the United States and Canada place their orders for two-wheel-drive, row-crop tractors and four-wheel-drive articulated units through their computer terminals. The orders are processed by two mainframe computers, IBM 3081 and IBM 3033, located in the Waterloo headquarters of the Tractor Works, which are the hosts for the four manufacturing facilities in Waterloo and Cedar Falls. There are ten DEC minicomputers. The structure is *hybrid*, the IBM and DEC computers working both in a horizontal and a hierarchical structure. In a horizontal structure, the computers are peers; in a hierarchical structure, the superior computer commands the inferior computer.

The computers constitute a *network* because they communicate with each other within a common context. They are *distributed* because hosts perform functions which affect the entire operations, such as the allocation of tractor demand to plants, while the slaves perform *local* functions, such as the PDP 11/03 microcomputer reporting the location of carts bearing reflective strips with binary coding to PDP 11/34 minicomputer controlling the load/unload stations.

Robot cranes pick up subassemblies from high-rise storage bins and deliver them to the assembly lines according to the master schedules worked out every night by the factory's master computer.

0.0 Computer-aided Design and Computer-aided Manufacturing

There are numerous terminals for computer-aided design (CAD) at all the Deere manufacturing facilities. These CAD terminals are directly linked to the NC and CNC controls on machine tools, permitting computer-aided manufacturing (CAM).

0.0 Group Technology

Group Technology (GT), specifically computer-aided GT, plays a significant role in the integrated manufacturing system at Deere:

> In GT, parts to be machined are grouped in families according to common characteristics. These include material, geometry, size, types of machining and volume. For each family of parts, Deere designs a set or "cell" of machine tools. These may be standalones or machining centers, and may be arranged along straight or curved paths. . . .[28]
>
> A system developed at John Deere has *saved the company $6 million in the last 18 months alone*, says Adel A. Zakaria, Deere's manager of manufacturing engineering systems and the system's chief architect. The system, called JD/GTS, contains *information on some 400,000 parts* and includes an English-like query language that enables rapid extraction and tabulation of parts information on a video screen. . . . The line's designers used JD/GTS to identify — among the thousands of types of gears used in Deere products — three families that could be made by the proposed facility. . . .
>
> On the heels of that success, the firm's component works is now installing a *flexible machining cell* for drilling tractor transmission blocks. Believed to be the *largest cell of its kind in the world*, it is comprised of *ten machining centers and two head indexers* — mammoth devices that change tools automatically under computer control. . . .
>
> Controlled by computers in a room overlooking the cell, the cell will handle eight different types of transmission blocks in as many as *20 different orientations* at a rate of some *300 units a day*.[29] (emphasis added)

0.0 CAD, CAM, and CAE for Computer-Integrated Manufacturing (CIM)

Notwithstanding the impressiveness of Deere Tractor Works operations, it is *not* an example of the automated factory. The manager of manufacturing and plant en-

28. Joe Quinlan, "The 'Waterloo Experience' and Deere's Philosophy," *Materials Handling Engineering*, June 1982, 82.

29. Paul Kinnucan, "Computer-aided Manufacturing Aims for Integration," *High Technology*, May–June 1982, 54.

gineering at Deere, T. E. Lorenzen, says: "We don't think of it as an automated factory. We talk about it as an integrated manufacturing system."[30]

Computer integrated manufacturing (CIM) is the term preferred by Richard L. Sikon, director of strategic product planning and market analysis of CAM Systems Business of Computervision Corporation:

> Any use of computer technology that addresses the overall objectives of a system which accepts as input product requirements and—through the use of various computerized approaches to the entire design through manufacturing cycle—delivers as output finished products ready for shipment.[31]

0.0 CAM—Direct and Indirect Applications

The basic integration is that of product requirements with manufacturing. We will define CAM, the manufacturing aspect, in the words of the not-for-profit research corporation, CAM-I:

> The effective utilization of computer technology in the management, control, and operations of the manufacturing facility through either direct or indirect computer interface with the physical and human resources of the company.[32]

CAM is directly applied to manufacture by monitoring and controlling the manufacturing operations. It is indirectly applied through production and inventory control, order scheduling, parts programming, etc.

0.0 Computer-aided Design—a Replacement for Drafting

The use of computers in manufacturing in the early 1970s was in testing finished electronic products. After automatic test equipment became well-accepted, the bottleneck moved upstream to manufacturing and layout. With the growing acceptance of computer-aided manufacturing, the bottleneck moved further upstream to design.

While the term CAD stands for computer-aided *design*, in practice, CAD tends to be computer-aided *drafting*. Arbab, et al. highlights the distinction and urges avoiding the creation of the gap in the first place:

30. Donald E. Hegland, "Putting the Automated Factory Together," *Production Engineering*, June 1982, 61.
31. *Ibid.*
32. Computer-Aided Manufacturing-International, Arlington, Texas.

> *Graphics* is the art of producing visual effects. . . .
> *Drafting* is the subfield of graphics where pictorial com-
> munication is through a special graphical language en-
> hanced by the mathematical transformations of projective
> geometry. . . . *Design*, on the other hand, is the process
> of planning and delineating an object, beginning with a
> designer's initial conception and ending with a variety of
> documents defining the object.[33]

0.0 Computer-aided Engineering (CAE)

One way of avoiding the creation of a gap between CAD and CAM is through
computer-aided engineering (CAE). Its contribution is in the early design phase,
from idea through prototype. Stephen Swirling and Gerard Langeler, of Mentor
Graphics, describe a *distributed* computer-aided design engineering system, which,
by reducing the number of secondary chores involved in design engineering, contributes to increased productivity. They show in figure 4.4 that the actual design
effort occurs only 58 percent of the design engineer's day, suggesting that productivity can be increased significantly by automating all aspects of his work that can
be taken over by a smart machine:

> The Idea 1000 package . . . built around Apollo Comput-
> er's small Domain computers and Aegis virtual memory
> operating system . . . has a 32-bit word width, virtual-
> memory management, a fast main memory (expandable
> to 3.5 Mbytes), a large disk-storage capacity, and local-
> area networking. . . .
>
> The Idea 1000 CAE system employs autonomous work
> stations that are connected by a 12-Mbit/s local-area net-
> work and supported by a *distributed data-base structure*.
>
> At each work station, a designer can break down a logic
> diagram, simulate circuits graphically (instead of with a
> breadboard), and change components and other circuit
> features. . . .
>
> With Idea 1000's *interactive logic simulation* software,
> called SIM, the designer in effect exercises his design be-
> fore he wire-raps a single pin.[34] (emphasis added)

33. Farhard Arbab, Larry Lichten, and Michael A. Melkanoff, "Toward CAM-oriented CAD,"
ACM IEEE Nineteenth Design Automation Conf. Proc., IEEE, 1982, 239–240.
   34. Stephen Swerling and Gerard H. Langeler, "CAE Tool Supports Both Design and Routine
Tasks," *Electronic Design*, 22 May 1982, 117,120.

While the CAE is discussed in the context of Integrated Circuit design, the essential element of relieving the design engineer of many chores holds a promise for any design effort, including that of design for assembly.

0.0 Toward the Automated Factory: Flexible Manufacturing System (FMS)

While product design for robomation may be far off, progress is possible in other elements of the operations converting inputs into outputs. As indicated in the definition of CIM above, the ideal use of computers would be in integrating market demand, on the one hand, with customer supply of products, on the other. This total span covering (market) demand and (manufacturers') supply is brought out by Wisnosky:

> [CIM is] the integration of the individual control elements. Functional areas, such as inventory control, physical distribution, cost accounting, and purchasing are integrated with the direct manufacturing control system.

Figure 4.4
Circuit Designer's Effort Distribution

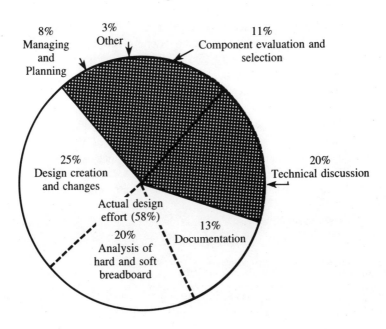

> Thus, the shop floor machines collect data for both manu-
> facturing and the overall organization. The machines also
> receive processing commands from the system.[35]

0.0 Flexible Line Assembly for Customization—
The Japanese Experience

By making the customer demand the prime mover, as reflected, in part, by the
change in inventory, the output is made responsive to the vagaries of demand.
Since it is possible to change the configuration of each output to meet customer
specifications and still move it through a robot assembly line, the demands for
customization can indeed be built into CIM. This is a feature that greatly im-
pressed Detroit auto executives when they visited the Zama auto plant of Japan's
Number 2 automaker, Nissan.

> Robots handle 97 percent of the work which is watched
> over by no more than half a dozen humans. It takes only
> one minute for each car to flow through the welding line.
> (*An entire vehicle is constructed in 18 hours.*)
>
> The tireless robots (each given the name of a Japanese fe-
> male pop singer to provide a slight human touch) allow
> Nissan to produce simultaneously the exact number of
> *tailor-made cars* required for *widely differing* foreign
> markets. . . .
>
> Assistant plant manager Teuro Osawa explains: "The ma-
> chine's sequence of movements is controlled by a central
> computer. Information is fed in telling the entire welding
> line they will be handling in order, say, a sedan, two
> hatchbacks, another sedan, a station wagon, and so
> on. . . .
>
> [The *flexible line assembly operation* permits] the Zama
> plant, for example, [to produce] an average of *50 cars a
> month* for the Danish market. In accordance with [the]
> Danish law, each *headlight* incorporates a small "wide-
> screen wiper," a feature not seen on any other of the
> thousands of vehicles pouring off the assembly line.[36]
> (emphasis added)

---

35. Dennis Wisnosky, "Computer Integrated Manufacturing: The Heart of the Factory of the Future," Speech delivered May 1981, cited in Harry Thompson and Michael Paris, "The Changing Face
of Manufacturing Technology," *Journal of Business Strategy*, Summer 1982, 47.

36. Geoffrey Murray, "How Nissan Wows U.S. Auto Execs," *The Christian Science Monitor*, 30
Sep. 1982, 1,9.

0.0 Flexible Manufacturing System (FMS)

The *flexible line assembly* at Nissan makes it possible for Japan's Number 2 auto-maker to meet even the exacting tailor-made demands, such as the fifty cars a month with widescreen wipers to meet Danish law requirements. To make it possible to meet such specialized requirements without disrupting the entire assembly line—as it does in Detroit—the manufacturing line has to be able to respond to the market demand rapidly and responsively. For instance, the manufacturing computer system should be able to recall from its database what the Danish "wide-screen wiper for left headlight" means in precise, executable instructions. Those specifications would have to be incorporated into the production schedule, so that the production schedule would have to specify that, say, on the 1,853rd unit on the assembly line, the widescreen wiper has to be installed. If for some reason a Nissan employee pulls a car off the assemblyline—and they are authorized to do so—, then the instructional sequence has to be changed, instructing that the widescreen wiper be installed on the 1,852nd unit on the assemblyline, instead of the 1,853rd.

The flexible line assembly is part of the flexible manufacturing system (FMS). The basic concept is that of computerized delivery synchronized with computerized monitoring, first introduced into the United States in 1972. The motivation for FMS comes in part from the excessive waiting time spent by each part awaiting its turn on the machine: as much as 95 percent of the time in the waiting line for 5 percent of the time on the machine.

While the CIM concept requires the integration of the entire process from the customer demand to the manufacturers' supply, FMS is the first step toward it. The FMS is a flexible concept itself, in the sense that under its rubric, a number of elements may be included: automated tool-changing capability associated with machines that can perform a number of different functions, flexible routing of a given part with the least delay from one machine to another, production scheduling, etc.

Irrespective of the formal definition of FMS, it is a working reality. Some 60 FMSs are in operation today: twenty-four in Japan, twenty-four in Western Europe, and twelve in the United States.[37] Figure 4.5 shows that FMS is appropriate in batch production of 1–1000 units per month.

0.0 FMS—the West German Experience

Of the sixty flexible manufacturing systems, perhaps the most sophisticated is that of Messerschmidt in Augsberg, West Germany. It has twenty-four machines, most of which have four or five axis heads. Earlier, in the discussion of Deere Tractor Works, we noted that Deere has a flexible machining cell comprising ten machine centers and two head indexers which change tools automatically under computer

37. Dennis Wisnosky, *op. cit.*

97

Robotics/Artificial Intelligence/Productivity

Figure 4.5
Relationship Between Automation and Number of Products

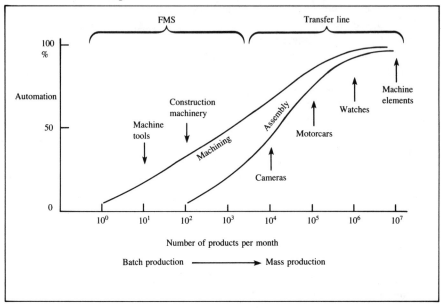

Source: Hiroyuki Yoshikawa, "State of the Art in Japan," in Committee on Computer-Aided Manufacturing, Manufacturing Studies Board, *Computer-Aided Manufacturing: An International Comparison*, National Academy of Sciences, Washington, D.C., 1981.

control. We also noted that Deere was able to reduce work-in-process inventory by 25 percent. GCA Corporation Group Vice President of Industrial Systems, Dennis Wisnosky, reports on the accomplishments at Messerschmidt:

> What has Messerschmidt gained from FMS? Machine utilization increased 44 percent, while the *number of machines decreased 44 percent*. In short, fixed assets are being used much better. The system decreased floor space used, *reduced the work force by 30 percent*, and *increased throughput by 25 percent*. The initial investment to install FMS was $50 million, which is actually 9 percent less the amount they would have been required for only select pieces of computer-controlled equipment. Considering all these factors, management claims *24 percent increase in annual returns*.[38] (emphasis added)

38. *Ibid.*

Figure 4.6
Integrated Monitoring and Control System

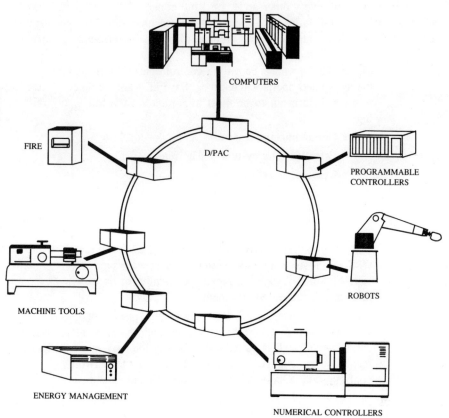

Whether viewed from the design viewpoint as CAD, the manufacturing viewpoint as CAM, or the integrative viewpoint as CIM, the key to robotics is control.

0.0 Monitoring and Control at Nissan in Smyrna, Tennessee
Nissan's brand new 3.2 million square foot, light truck factory in Smyrna, Tenn., is run by its automated control system known as NEMAC (Nissan Energy Management/Maintenance/Monitoring and Communications/Control). Ed Miller, President of the Oak Ridge Facility and Manufacturing Automation (ORFMA), which supplies NEMAC, says that a properly-designed, fully-integrated monitoring and control system of the type illustrated in figure 4.6, can easily result in *cost savings of 40 percent or more.*[39]

39. Ed Miller, "Technology and Automation: The Nissan Experience," *Survey of Business*, Summer 1982, 26.

Robotics/Artificial Intelligence/Productivity

Miller urges that the choice of communications network be given priority in integrating manufacturing operations. It should provide the ability for each node to communicate with every other node, i.e., full connectivity, across the different protocols of communications conventions of the machine tools, maintenance systems, programmable controllers and computers from many different vendors:

> The focus then must be on the control console itself and the associated operator interfaces. Several of the parameters which turn out to be most important are the following:
> - Communications speed
> - Effective total system update rate
> - Operator interfaces
> - Menu-driven page design
> - Resolution
> - Cost. . . .
>
> The state-of-the-art network system designed for Nissan, OPTONET, is a modular system of hardware and software building blocks developed specifically for operation in a manufacturing environment. . . .
>
> The software has three different logical aspects: an operating system, database management, and various application modules. It is structured to permit the easy implementation of changes or enhancements to the network configuration. Operators interact through fail-safe menu selections, and the system provides extensive self-diagnostics at all levels.[40]

The subject of controls being at the very heart of robotics, we will now turn to computer control of robotic manipulators.

40. *Ibid.*

5

Implementing Integrated Controls—Hardware and Software

OVERVIEW: Robot manipulation, including robot vision, could be considered control *of* the metal, while highly parallel computing *à la* the Fifth-Generation Computing System could be considered control *in* the metal.

The robot comprises essentially three components: a mechanical unit consisting of rigid bodies (links) connected by joints, a power supply, and a controller. Each joint-link pair constitutes a degree of freedom. To pick up objects and place them, the robot arm has to respond to forces applied by its actuators. The force is transmitted through the joint-link chain. The motion dyanmics are formulated in three different ways, each of which offers computational efficiency, structural accuracy, and real-time control in varying degrees. Conventional servomechanisms exercise control via continuous feedback on the actuators. More sophisticated controls permit the user to specify the direction and speed of the manipulator along a single straight line. Some even provide adaptive control, compensating for a wide range of manipulator motion and payloads.

Robot dynamics and control are accomplished through commands which the robot "understands." A comparative study of high-level robot languages in terms of eleven characteristics shows AUTOPASS as the most programmable language, but it requires complicated support facilities presently unavailable. The two most important aspects of developing a robot language are modular, expandable structure, and the ability to interact with external devices and sensors.

One of the most effective ways that a robot can interact with external devices and sensors is through robot vision. The process of computer vision for robot control, as well as for automatic visual inspection, can be divided into five principal areas: *sensing, segmentation, description, recognition,* and *interpretation.* The imbedding of controls in the machine, as in machine vision, attempts to make hardware progressively take over the software functions.

With respect to electronic functions, the description of "more for less" ap-

101

plies. Robert Noyes, the founder of the most famous of all microprocessor manufacturers, Intel Corporation, says that the cost per bit (binary digit) of random-access memory has declined an average of 35% per year since 1970. He predicts that by 1986, the cost per electronic function will have declined to 5% of the 1976 cost. What are the parameters of the electronic function? We identify four: transit time, clock time, length unit, and switching energy.

How do the various chip architectures rank with respect to their ability to balance the system parameters? John E. Price, of Amdahl, finds the gate array logic the overall winner. While the gate array is useful when computations call for identical arrangement of logic functions, programmable logic arrays (PLAs) are needed when the logic functions cannot be implemented by means of the regular combinational functions of the gate array. IBM has found that the laser offers 7.25 man-years savings over conventional design approaches. This and other elements of progress in VLSI technology make realizable the goal of highly parallel computing, wherein hundreds of thousands of computing elements can work cooperatively to solve a single problem. Ullman's discussion offers the most recent algorithms and results in "Ideal Parallel Computers and Their Simulation."

0.0 Communication Software for Hardware Controls

At the heart of successful automation of manufacturing is the control exercised through communication. Effective communication requires accurate concepts. After a decade of effort in assembly research, the director of the robotics research center at Draper Laboratories says: "You can't automate something you don't understand," referring to parts-mating, parts assembly, etc. More than the mechanistic descriptors of the physical process, it is the mental flow-charting performed by the human operators, routinely and repetitively, that has to be thoroughly understood before they can be converted into humanspeak, which in turn is conveyed through the abstract machinespeak of programs, which in turn is converted into the blips and beeps of direct machinespeak, which in turn accomplishes the pick-and-place, stamp-and-form, and other operations that the robotic arm can perform.

We will first discuss robot manipulators, including robot arm dynamics and control, and robotics software. Then we will review robot vision, including pattern recognition methods for vision and sensing for robot control. The imbedding of controls in the machine, as in machine vision, attempts to make the hardware take over software functions, an effort made possible by advances in fabricating integrated circuits. The progress and prospects of very large scale integration (VLSI) are discussed in the concluding section.

0.0 Robot Manipulators

The earliest stimulus to develop robot manipulators came from hazardous materials handling, in particular, radioactive material handling. In 1947, Argonne Na-

tional Laboratory started work on remote manipulators, in which the "slave" would duplicate the human hand motions of the "master," thus permitting radioactive materials to be handled without danger to humans.

The search to devise robots which could work without the constant supervision required in the "master-slave" environment led Unimation Inc. to develop the first commercially available industrial robot in 1959. Two years later, Ernst at MIT developed a computer-controlled robot with tactile sensors which could "feel" the blocks it was stacking up.[1] In the following year, Tomovic and Boni developed a robot with a pressure sensor which "felt" the object and sent a feedback signal to a motor which would grasp the object.[2]

In 1968, McCarthy and colleagues at Stanford developed "a computer with hands, eyes and ears."[3] Subsequent work at Stanford by Bolles and Paul incorporated both visual and force feedback into robotics manipulation in 1972.[4] In 1974, IBM programmed a robot to assemble twenty-two of the twenty-five pieces of the "rail support" in an IBM typewriter.[5] In the process of that demonstration, they investigated the problem of orienting mechanical parts,[6] and invented a vise[7] that simplified the reorientation of parts. Feedback from the robot's environment via different types of sensors began to receive increasing attention.[8]

0.0 Robot Arm Dynamics and Control

The robot comprises essentially three components: a mechanical unit consisting of rigid bodies (links) connected by joints, a power supply, and a controller. One end of the mechanical unit is attached to a base, while the other end is free to manipulate objects or perform assembly tasks. The complexity of the mechanical operations depends on the complexity of the control.

What is a robotic manipulator? It may be defined as a multidegree of freedom open-loop chain of mechanical linkages and joints, . . . driven by actuators, . . .

1. H. A. Ernst, "MH-1, A Computer-oriented Mechanical Hand," in *Proceedings of 1962 Joint Computer Conference* (San Francisco, Cal.: 1962), 39–51.

2. R. Tomovic and G. Boni, "An Adaptive Artificial Hand," *IRE Transactions on Automatic Control*, Vol. AC–7, No. 3, 1962, 3–10.

3. J. McCarthy, et al., "A Computer with Hands, Eyes, and Ears," *1968 Fall Joint Computer Conference AFIPS Proceedings* (1968), 329–338.

4. R. Bolles and R. Paul, "The Use of Sensory Feedback in a Programmable Assembly System," *Stanford Artificial Intelligence Laboratory Memo* AIM-220 (Stanford: Stanford Univ, 1973).

5. P. Will and D. Grossman, "An Experimental System for Computer Controlled Mechanical Assembly," *IEEE Transactions on Computer*, Vol. C-24, No. 9, 1975.

6. D. Grossman and M. Blasgen, "Orienting Mechanical Parts by Computer-controlled Manipulator," *IEEE Transactions on Systems, Man, and Cybernetics*, Vol. SMC-5, 1975, 561.

7. J. Griffith, D. Grossman, P. Will, and R. Garrison, "Quasi-liquid Vise for a Computer-controlled Manipulator," IBM Research Report RC-5451, (Yorktown Heights, New York: IBM T.J. Watson Research Center, 1975).

8. S. Wang and P. Will, "Sensors for Computer Controlled Mechanical Assembly," *Industrial Robot*, Vol. 4, No. 9, 1978.

Robotics/Artificial Intelligence/Productivity

Figure 5.1
Examples of Industrial Robots

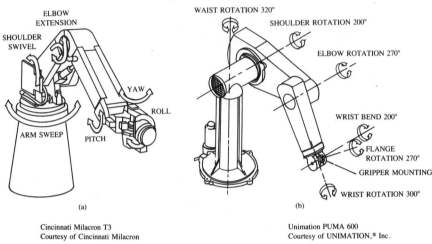

(a)

(b)

Cincinnati Milacron T3
Courtesy of Cincinnati Milacron

Unimation PUMA 600
Courtesy of UNIMATION,® Inc.

Source: *IEEE Transactions on Automatic Control*, Vol. AC-28, No. 2, Feb. 1983, 8

and capable of moving an object from initial to final locations or along prescribed trajectories.[9]

Each joint-link pair constitutes a degree of freedom. A typical industrial robot has six joints, providing it with six degrees of freedom. These are identified in the three arm movements and the three wrist movements of Cincinnati Milacron T3 in figure 5.1. The rotary axis movements of Unimation PUMA 600 are also shown.

These degrees of freedom are classifiable into four basic categories: (i) cartesian coordinate (three linear axes), (ii) cylindrical coordinate (two linear and one rotary axes), (iii) spherical or polar coordinate (one linear and two rotary axes), and (iv) revolute or articulated coordinate (three rotary axes).[10] The four categories are illustrated in figure 5.2.

To pick up objects and place them, the robot arm has to respond to force applied by its actuators. The force is transmitted through the joint-link chain. The motion dynamics are formulated in three different ways: N-E (Newton-Euler), L-E (Lagrange-Euler), and G-D (Generalized d'Alembert).

In N-E, the torque (total vector moment, about the center of mass is related to the angular velocity and angular acceleration of the body of Euler's equation:

9. David D. Ardayfio and Hardy J. Pottinger, "Computer Control of Robotic Manipulators," *Mechanical Engineering*, August, 1982, 40.

10. C. S. George Lee, "Robot Arm Kinematics, Dynamics, and Control," *Computer*, Dec. 1982, 62.

104

Figure 5.2
Various Robot Arm Categories

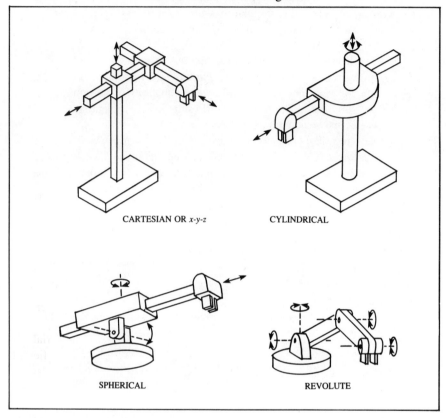

Source: C. S. George Lee, "Robot Arm Kinematics, Dynamics, and Control," *Computer*, December 1982, 63.

$$N = I \cdot w + w \ x \ (I \cdot w)$$

where w is the angular velocity and I is the inertia tensor.[11] The N-E method generates an efficient set of recursive equations. Hollerbach realized that the efficient recursive nature of N-E could be achieved with the Langrangian formulation as well.[12] The L-E approach can be expressed in matrix notation. Silver shows that there is no fundamental difference in computational efficiency between the two.[13]

11. K. R. Symon, *Mechanics*, (Reading, Mass: Addison-Wesley, 3rd ed., 1971).

12. J. M. Hollerbach, "A Recursive Langrangian Formulation of Manipulator Dynamics and a Comparative Study of Dynamic Formulation Complexity," *IEEE Transactions on Systems, Man and Cybernetics*, Nov. 1980, 730–736.

13. William M. Silver, "On the Equivalence of Langrangian and Newton-Euler Dynamics for Manipulators," *The International Journal of Robotics Research*, Vol. 1, No. 2, Summer 1982, 118.

Robotics/Artificial Intelligence/Productivity

In 1971, Chace and Bayazitoglu developed and applied a generalized d'Alembert force for multifreedom mechanical systems.[14] Lee, et al., extended Chace's work to mechanical manipulators with rotary joints.[15] Subsequently, Lee compared the three formulations in terms of computations as shown in table 5.1:

> The L-E approach is well structured and can be expressed in matrix notation but is *computationally impossible to use for real-time control* unless the equations of motion are simplified. The N-E method results in an efficient set of recursive equations, but they are difficult to use for deriving advanced control laws. The G-D equations of motion give fairly "well-structured" equations at the expense of higher computations. In short, a user can choose a formulation that is highly structured but *computationally inefficient (L-E)*, a formulation that has *efficient computations* at the expense of the motion equation structure *(N-E)*, or a formulation that retains the structure of the problem with only a *moderate computing penalty (G-D)*.[16] (emphasis added)

The inaccuracies arise from ignoring the nonlinearities in the motion dynamics of a manipulator with n degrees of freedom. It takes about eight seconds on a Fortran simulator to compute the move between two adjacent points on a planned trajectory for a six-joint robot, such as the Stanford arm. To expedite the computations, second-order terms such as the coupling reaction forces between joints— Coriolis and centrifugal forces—have been neglected.[17,18] These approximations adversely affect the performance at both low and high speeds. At low speeds, the arm movements are restricted to low speeds; at high speeds, the accurate position control of the robot arm becomes virtually impossible.[19]

Conventional servomechanisms are most often used to control robotic manipulators via feedback on the joints or actuators of the manipulator, which continuously measures the position of each axis. Several control methods are avail-

14. M. A. Chace and Y. O. Bayazitoglu, "Development and Application of a Generalized d'Alembert Force for Multifreedom Mechanical Systems," Transactions of ASME, *Journal of Engineering for Industry*, Series B, Vol. 93, Feb. 1971, 317–327.

15. C. S. G. Lee, B. H. Lee, and R. Nigam, "Development of the Generalized d'Alembert Equations of Motion for Mechanical Manipulators," *Proceedings of the 22nd Conference on Decision and Control*, IEEE, 1983.

16. C. S. G. Lee, "Robert Arm Kinematics," *op. cit.*, 77.

17. A. K. Bejczy, "Robot Arm Dynamics and Control," Tech. Memo 33-669 (Pasadena, Cal.: Jet Propulsion Laboratory, Feb. 1974).

18. R. P. Paul, *Robot Manipulators—Mathematics, Programming and Control* (Cambridge, Mass.: MIT Press, 1981).

19. C. S. G. Lee and M. J. Chung, "An Adaptive Control Strategy for Computer-based Manipulators," *Proc. 22nd Conf. on Decision and Control*, IEEE, 1983.

Table 5.1

Comparison of robot arm dynamics formulations

(n = number of degrees of freedom of the robot arm).

Approach	Lagrange-Euler	Newton-Euler	Generalized d'Alembert
Multiplications	$\dfrac{128}{3}n^4 + \dfrac{512}{3}n^3$ $\dfrac{739}{3}n^2 + \dfrac{160}{3}n$	$132n$	$n(\dfrac{31}{2}n^2 + 44n + \dfrac{63}{2})$
Additions	$\dfrac{98}{3}n^4 + \dfrac{781}{6}n^3$ $\dfrac{559}{3}n^2 + \dfrac{245}{6}n$	$111n - 4$	$n(9n^2 + \dfrac{69}{2}n + \dfrac{53}{2})$
Kinematics Representations	4 + 4 Homogeneous Matrices	Rotation Matrices and Position Vectors	Rotation Matrices and Position Vectors
Equations of Motion	Closed-form Differential Equations	Recursive Equations	Closed-form Differential Equations

Source: C. S. George Lee, "Robot Arm Kinematics, Dynamics, and Control," *Computer*, Dec. 1982, 78.

able.[20-34] While the servo-control method provides no information on the future path of the manipulator, the feedforward, or preview control method does use

20. A. K. Bejczy, *op. cit.*

21. R. P. Paul, "Modeling, Trajectory Calculation, and Servoing of a Computer Controlled Arm," Stanford Artificial Intelligence Laboratory Memo AM-177, Nov. 1972.

22. C. S. G. Lee, et al., "On the Control of Mechanical Manipulators," *Proc. Sixth Conf. on Estimation and Parameter Identification*, June 1982, 1454–1459.

23. D. E. Whitney, "Resolved Motion Rate Control of Manipulators and Human Prostheses," *IEEE Trans. Man-Machine Systems*, Vol. MMS-10, No. 2, June 1969, 47–53.

24. M. E. Kahn and B. Roth, "The Near-Minimum-Time Control of Open-Loop Articulated Kinematic Chains," *Trans. ASME, Journal of Dynamic Systems, Measurement, and Control*, Vol. 93, Sept. 1971, 164–172.

25. J. S. Albus, "A New Approach to Manipulator Control: The Cerebellar Model Articulation Controller (CMAC)," *Trans. ASME, J. Dynamic Systems, Measurement, and Control*, Vol. 97, Sept. 1975, 220–227.

26. H. Hemani and P. C. Camana, "Nonlinear Feedback in Simple Locomotion System," *IEEE Trans. Automatic Control*, Vol. AC-19, Dec. 1976, 855–860.

27. G. N. Saridis and H. E. Stephanou, "A Hierarchical Approach to the Control of a Prosthetic Arm," *IEEE Trans. Systems, Man, and Cybernetics*, Vol. SMC-7, No. 6, 1977, 407–420.

28. K. K. D. Young, "Controller Design for a Manipulator Using Theory of Variable Structure Systems," *IEEE Trans. Systems, Man, and Cybernetics*, Vol. SMC-8, No. 2, Feb. 1978, 101–109.

29. G. N. Saridis and C. S. G. Lee, "An Approximation Theory of Optimal Control for Trainable Manipulators," *IEEE Trans. Systems, Man, and Cybernetics*, Vol. SMC-9, No. 3, Mar. 1979, 152–159.

30. S. Dubowsky and D. T. DesForges, "The Application of Model Referenced Adaptive Control

information on how the path changes immediately ahead. Both feedback and feedforward components are incorporated in the *computer torque technique* based on L-E or N-E equations of motion.[35-36] The feedforward compensates for the interactions among various joints, while the feedback compensates for deviations from the ideal trajectory.

Resolved Motion Rate Control (RMRC) permits the user to specify the direction and speed of the manipulator along a single straight line. The motions of the various motors are combined and resolved into separately controllable hand motions along the world coordinate system. Whitney states the purpose:

> The objective of coordinated control is to allow the operator of a mechanical arm to command rates of the arm's hand along coordinate axes which are convenient and visible to the operator. This is a distinct improvement over conventional rate of control, which forces the operator to command the individual joint angle rates one at a time, often by means of an array of switches. A useful set of coordinates for hand control would provide motion along axes fixed to the hand itself, as shown in Fig. [5.3].
>
> To accomplish such motions, several joints of the arm must move simultaneously at time-varying rates. It is extremely difficult to accomplish this if conventional rate control is used. Some means of coordinating the joint motions is needed. Stated differently, the problem is to resolve the useful command directions into the necessary joint motions. For this reason, the method presented here is called resolved motion rate control.[37]

of Robotic Manipulators," *Trans. ASME, J. Dynamic Systems, Measurement, and Control*, Vol. 101, Sept. 1979, 193–200.

31. J. Y. S. Luh, M. W. Walker, and R. P. Paul, "Resolved-Acceleration Control of Mechanical Manipulators," *IEEE Trans. Automatic Control*, Vol. AC-25, No. 3, June 1980, 468–474.

32. R. Horowitz and M. Tomizuku, "An Adaptive Control Scheme for Mechanical Manipulators—Compensation of Nonlinearity and Decoupling Control," paper presented at the ASME Winter Annual Meeting, Nov. 1980.

33. A. J. Kovio and T. H. Guo, "Control of Robotic Manipulator with Adaptive Controller," *Proc. 20th IEEE Conf. Decision and Control*, Dec. 1981, 271–276.

34. C. S. G. Lee and M. J. Chung, "An Adaptive Control Strategy for Computer-Based Manipulators," *Proc. 21st IEEE Conf. Decision and Control*, Dec. 1982.

35. B. R. Markiewicz, "Analysis of Computed Torque Drive Method and Comparison with Conventional Position Servo for a Computer-controlled Manipulator," Technical Memorandum 33-601 (Pasadena, Cal.: Jet Propulsion Laboratory, 1973).

36. A. K. Bejczy, "Robot Arm Dynamics and Control," Tech. Memo 33-669 (Pasadena, Cal.: Jet Propulsion Laboratory, 1974).

37. D. E. Whitney, "The Mathematics of Coordinated Control of Prosthetic Arms and Manipulators," American Society of Mechanical Engineers Paper No. 72-WA/Aug. 4, '72.

Figure 5.3
Hand with Hand-oriented Coordinate System

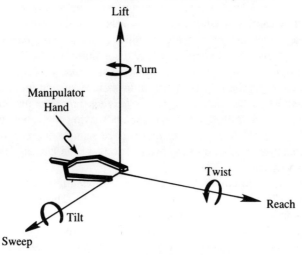

Source: D. E. Whitney, "The Mathematics of Coordinated Control of Prosthetic Arms and Manipulators," *Journal of Dynamic Systems, Measurement and Control*, Dec. 1972, 188.

While RMRC and other control methods control the robot arm at the hand or the joint level, their feedback strategies may become ineffective owing to their neglect of the changes of the load in the task cycle. Compensating for a wide range of manipulator motion and payloads is the concern of adaptive control techniques, among which the most widely used is Model-Referenced Adaptive Control (MRAC):

> A linear second-order, time-invariant, differential equation is selected as the referenced model for each degree of freedom of the robot arm. The manipulator is controlled by adjusting the position and velocity feedback gains to follow the model so that its closed-loop performance characteristics closely match the set of desired performance characteristics of the referenced model. As a result, the above simple MRAC only requires moderate computations to compute the control effort; this can be done relatively easily on a low-cost microprocessor.[38]

38. C. S. G. Lee, R. C. Gonzalez, and K. S. Fu, *Tutorial on Robotics*, (Los Angeles, Cal.: IEEE Computer Society, 1983), 178.

0.0 Robotics Software

The robot dynamics and control are accomplished through commands which the robot "understands." The commands have to be communicated by the user in such a way that they can be translated into a language that the robot "understands."

One way to communicate with the robot is via *discrete phrase recognition.* The state-of-the-art is rather primitive: the recognition is of a limited vocabulary, stored in a large memory space. The speaker has to pause between words. Using discrete words to get the robot to perform tasks is at best an arduous process.

A second way is to teach the robot by *leading it through* the operations using manual control, then editing the playback, and having the robot run through the motions using the playback. The memory-space required is relatively small; and the user learns to move the joystick, push-buttons, or the master-slave manipulator system relatively easily.

A third way is to employ a *high-level programming language,* or robot control language (RCL), for man-robot communication. The programming may be explicit or implicit. In explicit programming, the user specifies all the sequences of the manipulator motion. In implicit programming, the user specifies task-oriented statements, instead of writing out all the motion operations.

Ardayfio and Pottinger provide a convenient classification of languages:

> Robot programming languages may be divided into three categories, according to the focus of the programmer.
>
> At the lowest level are the *Joint Model languages.* Here the programmer's attention is focused on the *control of individual joints,* actuators, and sensors. ML, as well as the most nonservo-controlled robot command languages, could be placed in this category.
>
> Most commercially available robot languages are in the category of *Manipulator Model languages.* Here the programmer's attention is focused at the end effector or manipulator tool tip, and the task is to *guide the tip from point to point* through cartesian coordinate space. Languages in this category include Unimation's VAL, Olivetti's Sigla, Bendix's Teach, and Phillips' INDA as well as several well-known languages in the research community such as MIT's WAVE.
>
> The third category of robot languages are the *World Model languages.* Here the programmer's job is more goaldirected. End effector motion is described in terms of the position and motion of the objects being manipulated. Thus the programming language can largely be done off-line in terms of a preestablished data base describing the

Figure 5.4
Robot Programming Languages Evolution and Classification

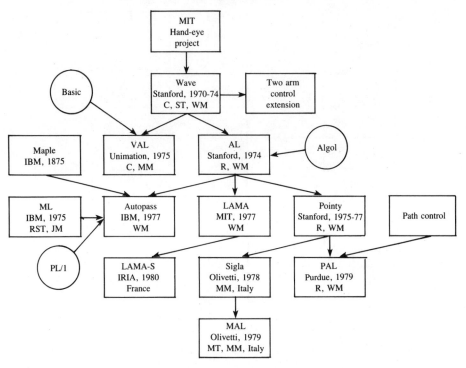

Source: David A. Ardayfio and Hardy J. Pottinger, "Computer Control of Robotic Manipulators," *Mechanical Engineering*, August 1982, 226.

parts to be assembled, or objects to be manipulated. This concept is more in line with an integrated CAD/CAM approach than the first two categories. Languages in this category include IBM's Autopass, Stanford's AL and LAMA, Purdue's PAL, and the University of Edinburgh's RAPT.

Figure [5.4] shows a sampling of some existing robot programming languages. They have been categorized according to the previous models, as well as whether they are used commercially or in a research environment (see Table [5.2]). As might be expected, the commercial languages tend to be more straightforward and conservative Manipulator Model types, whereas most research interest is in the World Model category. Where it is available, evolutionary history and the influence existing high-level

Table 5.2

Classification of Robot Language							
Robot Language	Commercial	Research	Single-task	Multitask	Joint model	Manipulator model	World model
Wave		*	*			*	
VAL	*					*	
AL		*					*
ML		*	*		*		
Autopass							*
Pointy		*					*
Sigla						*	
MAL				*		*	
PAL		*					*
LAMA							*
LM		*				*	
Robex		*				*	
Inda	*		*			*	
Teach	*					*	
Help	*					*	
Rail	*					*	

Source: David A. Ardayfio and Hardy J. Pottinger, "Computer Control of Robotic Manipulation," *Mechanical Engineering*, August 1982, 45.

languages have had on robot programming, has been included.

Summary information for each language includes the name (e.g. VAL), date it first appears in the literature (e.g., 1975), developing organization (e.g., Unimation), and whether it is intended mainly for commercial (C) or research (R) use, whether it is considered a single task (ST) or multitask (MT) language, and the type of coordinate model on which the language is based (WM—world model; MM—manipulator model; JM—joint model).[39] (emphasis added)

IBM considers their AML, A Manipulator Language, developed in 1977 by Russ Taylor, Phil Summers, and Jenine Meyer, a *second-generation* research robot system. Taylor points out that AML allows the user to combine manipulation, sensing, computational, and data processing functions provided by the system. Figure 5.5 presents the major functional components of the integrated robot system. There are three major software components. The first, AML, provides an *interactive* programming environment for application programming. The second performs trajectory planning, motion coordination, sensor monitoring, and other real-time activity. The third performs supervisor services, such as file and terminal I/O. During the four years, the system has been used in assembly, inspection and test, light fabrication, and integrated materials handling. Taylor reports on the flexibility offered by the combination:

> The flexibility inherent in this combination offers important advantages over current fixed automation. Viewed as a *stand-alone device,* a robot can use sensing and software to compensate for positional misalignments, reducing the requirement for high accuracy and expensive fixturing. Viewed as a part of a *horizontally integrated manufacturing system,* a robot can accomodate varying workpieces, spreading the capital costs over more products. It can interleave more efficient production scheduling. Viewed as part of a *vertically integrated design and manufacturing system,* a robot can allow the automated manufacturing of unique customized products.[40] (emphasis added)

39. David A. Ardayfio and Hardy J. Pottinger, "Computer Control of Robotic Manipulators," *Mechanical Engineering*, August 1982, 45.
40. R. H. Taylor, "An Integrated Robot System Architecture," *IBM Research Report*, (Yorktown Heights, N. Y.: IBM Research Center, Jan. 1983), 1.

Robotics/Artificial Intelligence/Productivity

Figure 5.5
Block Diagram of Autopass System

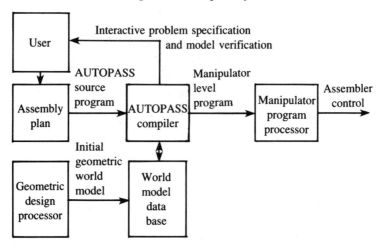

Source: "AUTOPASS: An Automatic Programming System for Computer Controlled Mechanical Assembly," *IBM Journal of Research and Development*, Vol. 21, No. 4, July 1977, 326.

Bonner and Shin made a comparative study of robot languages. They expanded on the six characteristics of a good programming language proposed by Pratt. To make a quantitative comparison of the fourteen robot languages they studied, they programmed the same example in all the languages:

> A robot is assigned to pick blocks off a conveyor belt and deposit them in a pallet with a three-by-three array of positions for the blocks. It is assumed that the blocks are precisely positioned on the conveyor and that the conveyor stops automatically when a block has arrived at the pickup point. Reactivation of the conveyor is done either manually or via signals from the robot program.[41]

Table 5.3 presents their results. AUTOPASS emerges as the most programmable language, but it also requires the most complicated support facilities which are presently undeveloped. They conclude that:

> [T]he most important aspects in the development of a robot language are the use of a modular, *expandable structure* and the ability to *interact with external devices and*

41. Susan Bonner and Kang G. Shin, "A Comparative Study of Robot Languages," *Computer*, Dec. 1982, 92.

114

Table 5.3

Quantitative comparison of the 14 languages for the palletizing-block programming example.

	Funky	T3	Anorad	Emily	RCL	RPL	Sigla	VAL	AL	Help	Maple	MCl	PAL	Autopass
Number of instructions	94	94	20	34 (27)	76	58	35	35 (22)	50	30	44 (35)	69	25	8
Development time	1,5	1,5	4	2	2,5	5,6	4	2	3	2	3	6	7	1
Understandability of instructions	2	2,6	6,7	3,4	3,4,5	3,4,5	6,7	3,4,5	1,4	3,4	1,4	3,4,5	3,6	1,4
Structured format	3	3	3	1	2	3	3	3	1	2	1	2	3	2
Flexibility of variables	5	4	2	2	2	2	4	2	1	1	1	2	3	1
Ease of extension	4	4	4	1	4	1	4	4	2	3	2	3	4	2
Range of users	1,2	1,2	2	2,3	3	4	2	2,3	4	2,3,4	3	4	5	1
Programming complex tasks	4	4	4	4	4	2	4	4	3 (1)	4	3	2	4	1

Table 5.3 (continued)

	Funky	T3	Anorad	Emily	RCL	RPL	Sigla	VAL	AL	Help	Maple	MCl	PAL	Autopass
Computing power	1	1	1	2	1	2	1	1	2	2	3	3	2	3
Sensing ability	2	1	1	2	1	3	1	1	1	1	2	3	1	4
Availability	2	1	1	2	2	2	1	1	2	1	3	1	2	3

Number of instructions. Of the different measures of programmability, the number of instructions in the program (excluding comments) is the only absolutely quantitative one. Where two numbers appear, the first represents the actual number of statements, and the second (in parentheses) is the number of statements the program can be reduced to when extraneous assignments are removed. (Assignment statements are added to increase readability.)

Development time. This measure is divided into seven aspects:

(1) fast and simple
(2) quick, but requires some thought and/or intelligence
(3) quick, with background in structured programming and knowledge of transforms
(4) quick, but instructions are not easy to read.
(5) encumbered by awkward control constructs
(6) encumbered by complicated coordinate transform usage
(7) requires extensive knowledge of coordinate transformation arithmetic.

Readability. Since readability is also quite varied and complex, it is divided into several submeasures that each affect the readability of the languages:

Ease of extension. Four degrees of expansion are possible among 14 languages:

(1) subroutines or other extension facilities available
(2) expandable through subroutines with some loss of English-like syntax
(3) somewhat expandable through subroutines but nesting is prohibited
(4) must rewrite code to add instructions.

Range of users. Five levels of experience describe the range of users:

(1) novice
(2) NC programmer or machine operator
(3) person with some programming experience
(4) person familiar with structured programming and/or transformations
(5) person with extensive knowledge of transformations.

Programming complex tasks. The languages vary in their ability to program different tasks. Some are limited to very simple pick-and-place operations while others are capable of complex tasks such as fastening and compliant motion. This measure is not directly related to the programming example, but is included for comparative purposes. The task abilities of the languages are divided into four levels:

These include

Understandability of instructions. Understandability can be characterized by seven aspects that relate to how the language reads.

(1) reads very much like English
(2) uses function button control
(3) instructions are words in English but do not read like English
(4) readability is markedly improved through proper choice of variable names
(5) instructions/variables limited to six characters
(6) instructions/variables limited to two or three characters
(7) instructions are not English-like at all.

Structured format. The degree of program structure in a language aids readability by enabling a known format for all programs and forcing variable declarations. There are three degrees of structure:

(1) structure is an inherent language feature
(2) can structure programs but not forced to do so
(3) the language is inherently unstructured.

Flexibility of choosing variables. Can the user define variable names? The following categories are considered:

(1) essentially unlimited variable naming
(2) variable names limited to six characters
(3) variable names limited to two or three characters
(4) variables available but only by number
(5) variables not available.

(1) capable of programming complex tasks using multiple arm operations.
(2) capable of programming complex tasks using visual feedback, such as part recognition
(3) capable of programming complex tasks using touch and force feedback sensing, such as compliant motion
(4) capable of only simple tasks with minimal or no touch sensing in fingers.

Necessary support facilities. Measures of support facility requirements, while not directly related to the programming example are included in the table for a more realistic portrayal of which languages can actually be used today. The measures of support considered are:

Computing power. The language requires a

(1) microcomputer or small minicomputer.
(2) minicomputer
(3) combination mainframe and small minicomputer.

Sensing ability. What robot capabilities are needed for the language's operation?

(1) only simple touch or no sensing required for complete operation
(2) proximity sensing specified in addition to touch sensing
(3) vision in the form of part recognition required
(4) complete dynamic world modeling system needed for operation.

Availability. Is the language as it is presented operational at present?

(1) operational and available commercially
(2) mostly operational but not commercially available
(3) only a small subset is operational.

Source: Susan Bonner and Karg G. Shin, "A Comparative Study of Robot Languages," *Computer*, Dec. 1982, 94.

sensors. Without modularity and expandability, a language will not only be difficult to interface to different systems but will soon become outdated because of advances in technology. Without interaction with sensors and peripheral devices, the robot will be isolated from the world around it and will never be able to perform with any high degree of proficiency. These two aspects are the key to the success of any robot programming language.[42] (emphasis added)

0.0 Robot Vision

One of the most effective ways that a robot can "interact with external devices and sensors" is robot vision. As a matter of fact, vision is indispensable for positioning and processing—positioning of objects and orienting them, and performing the many inspections while processing that are done implicitly by humans. Without vision, the parts and tools must be positioned precisely for the robot and oriented properly to perform the assigned tasks. And the special-purpose feeding and positioning mechanisms are quite expensive. To approximate the implicit in-process inspection that humans do, extensive specifications have to be made against which comparisons have to be performed. Both these disadvantages may be mitigated by machine vision.

0.0 Areas of Needed Research

In May 1982, the IEEE Computer Society sponsored a three-day workshop on Industrial Applications of Machine Vision at Research Triangle Park in North Carolina, with sixty-five researchers from nine countries participating. John F. Jarvis, head of the Robotics Systems Research Department at Bell Laboratories, provided the workshop summary, reviewing the state-of-the-art and identifying the areas of needed research:

> One can define robot vision as a means of *dealing with the unexpected*. In the context of robotics, the unexpected is primarily uncertainty regarding the *position of the object or the robot*. The issue of why location and orientation must be recovered arose in the workshop. Why can't position and orientation be preserved? Such preservation may be too costly because of the need for special fixtures or parts carriers. . . .
>
> It appears that all techniques of computer vision are potentially applicable to robot vision. . . .

42. *Ibid*. 95.

Areas of particular relevance to robotics include coordination of multiple sensors, two or more vision centers, or
vision and tactile sensors. Since system speed and cost
are intimately related, progress in implementing custom
VLSI will allow a radical increase in the amount of processing that can be applied to a given task. Specific attention must also be given to programming languages for
dealing with sensory data represented as two- or three-
dimensional arrays. In addition, study is still required to
determine exactly what primitives, data structures, and
control structures are most appropriate for specifying vision algorithms. . . .

The outlook for robot vision is excellent.[43] (emphasis
added)

0.0 Pattern-Recognition Methods for Vision

Jarvis's characterization of robot vision as a means for dealing with the unexpected underscores the requirements of discrimination in decision-making. Vision
in general, and visual inspection in particular, can be viewed as pattern-recognition, the methods for which are grouped as follows: decision-theoretic or discriminant approach,[44-52] template-matching approach,[53] and syntactic and structural
approach.[54-56] The template-matching is a special case of the decision-theoretic
approach. While the decision-theoretic approach requires representation of N fea-

43. John F. Jarvis, "Research Directions in Industrial Machine Vision: A Workshop Summary,"
Computer, Dec. 1982, 57–58.

44. J. R. Ullman, *Pattern Recognition Techniques*, Crane, Russak & Co., N.Y. 1973.

45. K. S. Fu, *Sequential Methods in Pattern Recognition and Machine Learning*, (New York:
Academic Press, 1968).

46. T. Cover, "Recent Books on Pattern Recognition," *IEEE Trans. Info. Theory*, Vol. IT-19,
Nov. 1973.

47. C. H. Chen, *Statistical Pattern Recognition* (New York: Hayden, 1973).

48. Y. T. Chien, *Interactive Pattern Recognition* (New York: Marcel Dekker, 1978).

49. R. O. Duda and P. E. Hart, *Pattern Classification and Scene Analysis* (New York: Wiley,
1973).

50. K. S. Fu, ed. *Digital Pattern Recognition, Communication and Cybernetics*, Vol. 10. (New
York: Springer, 1976).

51. J. T. Tou and R. C. Gonzalez, *Pattern Recognition Principles* (Reading, Mass.: Addison-
Wesley, 1974).

52. K. S. Fu and A. Rosenfeld, "Pattern Recognition and Image Processing," *IEEE Trans. Computers*, Vol. C-25, No. 12, Dec. 1976, 1336–1346.

53. K. S. Fu, ed. *IEEE Proc.* Special Issue on Pattern Recognition and Image Processing, May
1979.

54. K. S. Fu, *Syntactic Pattern Recognition and Applications* (Englewood Cliffs, N.J.: Prentice-
Hall, 1982).

55. T. Pavlidis, *Structural Pattern Recognition* (New York: Springer, 1977).

56. R. C. Gonzalez and M. G. Thomason, *Syntactic Pattern Recognition: An Introduction* (Reading, Mass.: Addison-Wesley, 1978).

tures of a pattern, and comparison of that N-dimensional vector with another vector to determine the distance measure, or a discriminant function, the template-matching requires a much simpler measure of similarity, such as correlation. Unlike the decision-theoretic approach, in the structural and syntactic approach, a pattern is usually represented as a string, or a tree, or a graph. Syntax analysis or parsing procedure determines the similarity between the two strings.

The process of computer vision for robot control, as well as for automatic visual inspection, can be divided into five principal areas: *sensing, segmentation, description, recognition,* and *interpretation.*[57]

0.0 Sensing for Robot Control

It will be recalled that Jarvis, in reporting on the areas of on-going research significant to robot vision, mentioned multiple sensors. The visual sensing is done by converting visual information into electrical signals by using visual sensors, preferably multiple sensors. While a single camera is most often used, two cameras are better than one to provide more information about the scene. Mese discusses the use of two cameras to achieve adequate resolution of LSI chip position.[58]

A typical vision system is presented in figure 5.6 from Trombly.[59]

Speed, resolution, and image contrast are major elements involved in the trade-off of a vision system. This picture signal from the camera that looks at the machine parts is divided into pixels (picture elements), ranging in size from 64×64 to 512×512 pixels, depending on the resolution requirements, which significantly influence the cost and speed. A small image window, approximately 128×128 pixels, takes about 0.2 s to process, while 320×240 would require 4 times as much at 0.8 s.

Trombly points out that area, position, number of holes, and part orientation can be calculated in about 0.25 s, for a rate of 240 objects/min. in a 128×128 pixel area. Four times as much time would be required for a 256×256 area, reducing the number of objects handled to 60/min. The typical maximum throughputs are between 60 to 300 objects/min.[60]

The processing of objects for vision and control purposes depends on detecting similarities and differences (which appear as discontinuities). In detecting a solid, such as inspecting a component on a conveyor, the discontinuities occur at the edges. Special lighting of edges would be helpful for edge detection. A system

57. R. C. Gonzalez and R. Safabaksh, "Computer Vision Techniques for Industrial Inspection and Robot Control: A Tutorial Overview," in C. S. G. Lee, R. C. Gonzalez and K. S. Fu, *Tutorial on Robotics* (Los Angeles, Cal.: IEEE Computer Society, 1983), 301.

58. M. Mese, I. Yamazaki, and T. Hamada, "An Automatic Position Recognition Technique for LSI Assembly," *Proc. Fifth Int. Jt. Conf. on Artificial Intelligence* (1977), 685–693.

59. John E. Trombly, "Adding Machine Vision to Assembly Lines," *Machine Design*, Nov. 11, 1982, 79.

60. *Ibid.*, 81.

Figure 5.6
Anatomy of a Vision System

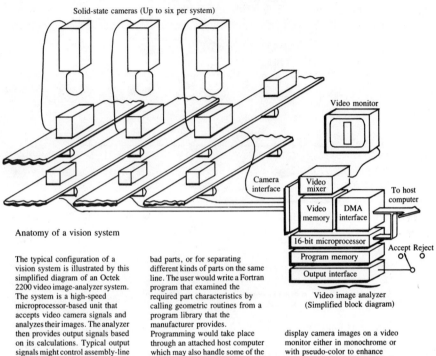

Solid-state cameras (Up to six per system)

Video monitor

Camera interface

Video mixer

To host computer

Video memory | DMA interface

16-bit microprocessor

Accept Reject

Program memory

Output interface

Video image analyzer
(Simplified block diagram)

Anatomy of a vision system

The typical configuration of a vision system is illustrated by this simplified diagram of an Octek 2200 video image-analyzer system. The system is a high-speed microprocessor-based unit that accepts video camera signals and analyzes their images. The analyzer then provides output signals based on its calculations. Typical output signals might control assembly-line hardware for separating good and bad parts, or for separating different kinds of parts on the same line. The user would write a Fortran program that examined the required part characteristics by calling geometric routines from a program library that the manufacturer provides. Programming would take place through an attached host computer which may also handle some of the image analysis. The system can display camera images on a video monitor either in monochrome or with pseudo-color to enhance differences in part contrast levels.

Source: John E. Trombly, "Adding Machine Vision to Assembly Lines," *Machine Design*, Nov. 11, 1982, 79.

can distinguish an edge if local background intensity is at level 3 on a scale of brightness from 1 to 10, and the edge is illuminated with at least 4.5. Nominal edge-level intensity should be 5.5 or greater.[61]

General Motors uses base lighting principle to inspect rough surfaces. Figure 5.7 shows the principle of their CONSIGHT-I system. A narrow, intense line of light, projected across a moving conveyor, is intercepted when there is an object on the conveyor, as indicated by the interference of the light in figure 5.7(b). To avoid the problem of shadows causing misreading of the object, two or more light sources are directed at the same strip of light, so that even if one source is prematurely interfered with, the other would not be. Figure 5.8 shows the monitor view of a part showing the interrupted light beams.

61. *Ibid.*

Figure 5.7
(a) Basic Lighting Principle Used in GM's Consight-I
(b) Computer View With and Without Parts

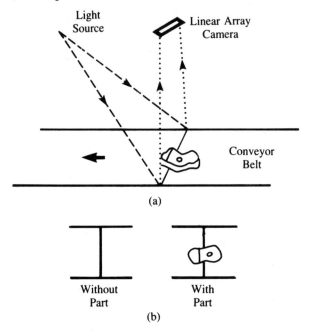

(a)

Without Part   With Part

(b)

Source: R. C. Gonzalez and R. Safabaksh, "Computer Vision Techniques for Industrial Inspection and Robot Control: A Tutorial Overview," in C. S. G. Lee, R. C. Gonzalez and K. S. Fu, *Tutorial on Robotics*, (Los Angeles, Cal.: IEEE Computer Society, 1983), 305.

0.0 Segmentation for Robot Control

Segmentation is the breaking up of the sensed area into its constituent parts or objects. Segmentation algorithms are generally based on either discontinuity (e.g., edge detection), or similarity (e.g., thresholding and region growing).

Edge detection is achieved by measuring the gradient in the x-direction and in the y-direction, at each pixel location, and using the gradient as the measure proportional to discontinuity in the image area examined under the mask.

Thresholding is the most widely used approach to segmentation. Proximity of points sharing a given property is used to separate objects from the background. Local thresholding is used when the objects are less clearly differentiated from the background; global thresholding is used when the objects vary markedly from the background with respect to some characteristic.

Region growing techniques are used when thresholding and edge detection do not apply. Shrinking, medial axis transformation, and region growing use the

122

value of a pixel as a function of the values of its eight nearest neighbors. Owing to its impracticality from the point of view of computation and/or hardware implementation, region growing is not widely used in industrial applications.

0.0 Description for Robot Control

Description is extracting enough features that are independent of object location and orientation to identify one object from another on the basis of discriminating information.

Shape, amplitude, perimeter, extreme of a region, etc., have been used as descriptors for industrial computer vision.

0.0 Recognition for Robot Control

Recognition is identifying each segmented object and assigning it a label.

There are two principal approaches: decision-theoretic and structural. The most often used technique of decision-theoretic recognition is matching. The most widely used technique of structural recognition is syntatic pattern recognition. While the decision-theoretic approach deals with pattern on a strictly quantitative basis, syntatic analysis deals with structural relationships ignored in the former.

Robotics/Artificial Intelligence/Productivity

0.0 Interpretation for Robot Control
Interpretation is endowing a vision system with a higher level of conception about its environment.

Low-level vision techniques extract "primitive information" from the observed scene, e.g., detecting the edge, while medium-level vision uses information from low-level vision to generate more meaning, e.g., linking of edges. Tsuji and Nakamura describe a system that can recognize unobscured parts of small internal combustion engines in a stack of parts.[62]

0.0 Imbedding Controls in Hardware
The imbedding of controls in the machine, as in machine vision, attempts to make hardware progressively take over software functions. We could sense some historic parallel here between the technology of hard-wired boards of the early computers and the subsequent technology of stored programs. The hard-wired boards could perform numerical calculations repeatedly and reliably; however, each type of calculations required its own wiring. Instead of physically changing the wired boards, stored programs in the computer could, as it were, change the required wiring—in fact, the sequence of data input, processing, and output—accomplishing different types of numerical (and later, symbolic) calculations. The world's first computations performed by a stored program were carried out in May 1949 at Cambridge University in England on the Electronic Delay Storage Automatic Calculator (EDSAC). The National Bureau of Standards followed in 1950 with the first American stored program loaded on the Standards Eastern Automatic Computer (SEAC).

Just about that time, revolutionary developments were taking place in microelectronics. Brattain and Boredeen invented the transistor in 1948 at Bell Telephone Laboratories. Patrick Haggerty of Texas Instruments foresaw the potentials of the new field:

> During 1949 and 1950 it finally became clear to me that the future of electronics would be profoundly influenced by knowledge already attained and additional knowledge being rapidly gained about materials at the structure-of-matter level. . . . In early 1951 we began to formalize our strategy by definite commitment to develop, manufacture and market semiconductor devices. . . .
>
> The three principal tactics we used to fulfill our semiconductor strategy were: (1) the development of the *first silicon transistor,* (2) the development and marketing of the

62. S. Tsuji and A. Nakamura, "Recognition of an Object in a Stack of Industrial Parts," *Proc. 4th Int. Jt. Conf. on Artificial Intelligence* (1975), 811–818.

> *first pocket radio,* the famous Regency, in collaboration
> with the IDEA Corporation, and (3) the development of
> our process for production of pure silicon. . . .
>
> From this *first semiconductor strategy,* the beginnings of
> a *second* can be traced to discussions in 1956 between
> Willis Adcock and me in which we *speculated on the fea-*
> *sibility* of whole circuits processed in minute wafers of
> pure silicon. The *dream* became a reality with the inven-
> tion of the *first practical integrated circuit* by Jack Kilby.
> . . . *Jack invented the integrated circuit in the summer of*
> *1958,* just ten years after the invention of the transistor.[63]
> (emphasis added)

The transistor is a low-amplifier which replaces the large, cumbersome,
power-hungry vacuum tube. Unlike the analog circuits which usually amplify ten-
fold each input, thereby reaching the practical limit of voltage levels with a few
microcircuits.

Once the transistor was found to successfully substitute for the amplification
of the vacuum tube, efforts were made to find other solid-state substitutes for
other electronic functions. In the integrated circuit, the body resistance of the
semiconductor itself and the capacitance of the junctions between positive (p) and
negative (n) regions are concentrated to realize a complete circuit of complex
combinations.

0.0 More for Less

The phenomenal growth in microelectronics continues today. Several statements
made by Robert Noyce, the founder of the most famous of all microprocessor
manufacturers, Intel Corporation, in a 1977 paper, are of historic interest and contemporary validity:

> Today's microcomputer, at a cost of perhaps $300, has
> more computing capacity than the first large electronic
> computer, ENIAC. It is twenty times faster, has a large
> memory, is thousands of times more reliable, consumes
> the power of a lightbulb rather than that of a locomotive,
> occupies 1/30,000 the volume and costs 1/10,000 as
> much. It is available by mail order or at your local hobby
> shop.
>
> In 1964, noting that since the production of the planar
> transistor in 1959 the number of elements in advanced in-

---

63. Patrick E. Haggerty, *Management Philosophies and Practices of Texas Instruments Incorpo-*
rated (Dallas, Texas: Texas Instruments, 1965), 50, 17.

Figure 5.9
Number of Components per Circuit

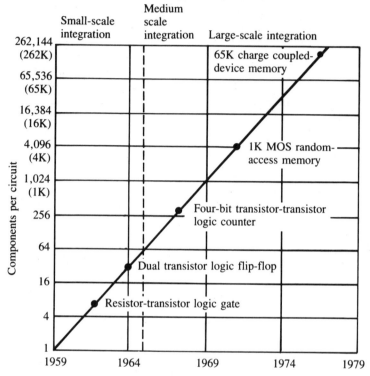

Source: Robert N. Noyce, "Microelectronics," *Scientific American*, Vol. 237, No. 3, Sept. 1977.

tegrated circuits had been doubling every year, Gordon E. Moore, who was then director of research at Fairchild, was the first to predict the future progress of the integrated circuit. He suggested that its *complexity would continue to double every year*. Today, with circuits containing $2^{18}[=262,144]$ elements available, we have not yet seen any significant departure from Moore's law. . . .

The most striking characteristic of the microelectronics industry has been a persistent and rapid decline in the cost of a given electronic function. . . . The cost of a given electronic function has been declining even more rapidly than the cost of integrated circuits, since the complexity of the circuits has been increasing as their price has decreased. For example, the *cost per bit (binary digit) of random-access memory has declined an average*

126

of thirty-five percent per year since 1970, when the major growth in the adoption of semiconductor memory elements got under way. The cost declines were accomplished not only by the traditional learning process but also by the integration of more bits into each integrated circuit: in 1970 a change was made from 256 bits to 1024 bits per circuit and now the number of bits is in the process of jumping from 4096 per circuit to 16,384. . . . [See figure 5.9]

In an industry whose *product declines in price by twenty-five percent a year* the motivation for doing research and development is clearly high. A year's advantage in introducing a new product or new process can give a company a twenty-five percent cost advantage over competing companies: conversely, a year's lag puts a company at a significant disadvantage with respect to its competitors. . . .

By 1986 the number of electronic functions incorporated into a wide range of products each year can be expected to be one hundred times greater than it is today. The experience curve predicts that the *cost per function will have declined by then to a twentieth of the 1976 cost, a reduction of twenty-five percent per year.*[64] [See figure 5.10] (emphasis added)

The change in performance/cost ratio of computers may be viewed historically. Turn, at the Rand Corporation, developed the original curve showing the cost in dollars per million instructions per second (MIPS) for general-purpose computers, reproduced as figure 5.11. Macintosh redrew Turn's curve to give the greatest weight to actual, well-documented cost data, and plotted the device price curve, as shown in figure 5.12. The curve dramatically depicts the steady drop in price with ever-increasing speed and reliability.

The projected benefits of advances in hardware technology for the command and control functions were the subject of a study by the U.S. Air Force. The forecast of hardware advances made by the Command and Control Information Processing in the 1980s, CCIP-85 Study, is reproduced as figure 5.13. We notice that IBM 360/195 appears in figure 5.11 as costing less than $1 million per MIP, with a capability of 10 MIPS as shown in figure 5.13. ILLIAC IV has ten times the capability of IBM 360/195, viz., 100 MIPS. The CCIP projects one hundred times the ILLIAC IV capability, i.e., 10,000 MIPS, to be achieved in 1990, while figure 5.11 places the timeframe around 1983, slightly over optimistic.

64. Robert N. Noyce, "Microelectronics," *Scientific American*, Vol. 237, No. 3, Sep. 1977.

Figure 5.10
Cost per Bit of Computer Memory

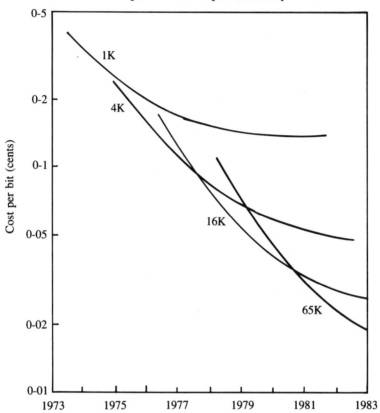

Source: Robert N. Noyce, "Microelectronics," *Scientific American*, Vol. 237, No. 3, Sept. 1977.

0.0 Basic Measures of Performance of Integrated Circuits

The ever-increasing speed and reliability levels, combined with ever-decreasing cost and size, that we associate with computing equipment, come from the advances in fabricating integrated circuits, in which we realize a complete circuit of resistors, capacitors, and amplifiers. Conducting materials are laid down on a wafer, usually of silicon. While a large-scale integrated circuit contains tens of thousands of elements, each circuit is less than a quarter of an inch on a side. Each circuit is made up of a number of layers. In the negative-channel, metal-oxide semiconductor (NMOS), field-effect transistor (FET), there are three layers: diffusion, polysilicon, and metal. When a path on the polysilicon crosses a path on the diffusion level, a *transistor* is created. The transistor is a simple switch: a voltage on the polysilicon-level path controls the flow of current in the diffusion-level

Figure 5.11
Cost Trends in High-Performance, General Application Computers

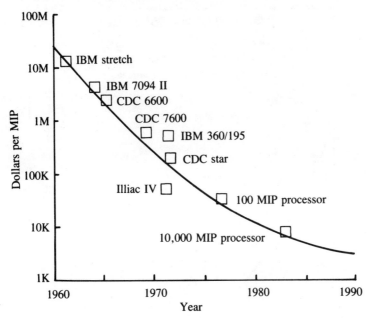

Source: "The Turn Report", the Rand Corporation.

path. The basic operation of the transistor is to use the charge on its *gate*—where the polysilicon crosses the diffusion—to control the movement of the negative charge between the source and the drain as shown in figure 5.14.

How do we measure the performance of the complete circuits?

0.0 Transit Time, τ

The objective of an integrated system is to transmit signals. The transit time τ is the average time required for an electron to move from source to drain. The lower limit of τ between two states of an electron is:

$$\tau \geq \Delta E$$

where h or h-bar $\simeq 1.05 \times 10^{-34}$ joule-second/radian. The transit time is the minimum time in which the charge placed on the gate of one transistor results in the transfer of a similar charge through the transistor's channel onto the gate of a subsequent transistor. In 1978, τ = 0.3 ns (nanoseconds).

What should τ be in supercomputers whose performance is measured in MIPS? Medrick calculates for a hypothetical 500 MIPS mainframe CPU, the cycle time τ_{cpu} = 660 ps, assuming 3 cycles per instruction. About 300K logic circuits

Robotics/Artificial Intelligence/Productivity

Figure 5.12
Comparison of Computing Costs and Silicon Device Costs

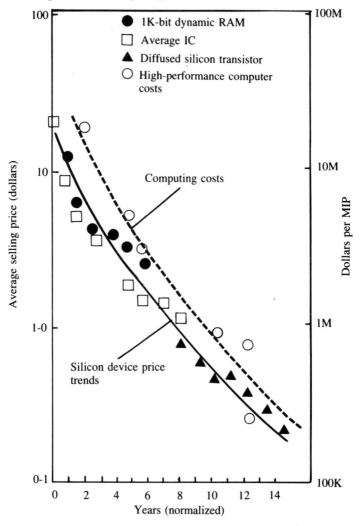

Source: Ian M. MacKintosh, "Micro: The Coming World War," *Microelectronics Journal*, Vol. 9, No. 2. 1978.

$(K = 2^{10} = 1,024)$ are needed, each with an average delay $\tau_l \simeq 1/10 \ \tau_{cpu} \simeq 66$ ps, and a cache capacity of 64 KB(Byte = 8 bits) at a cycle time and an access time $\tau_{ca} = 1/3 \ \tau_{epu} \simeq 220$ ps. The main memory capacity is 250–500 MB at a cycle time $\gamma_m \simeq 6.6$ ns.[65]

65. R. E. Medrick, *Computer Storage Systems and Technology*, Wiley, N.Y., 1977.

Figure 5.13
Computing Power Requirements Technology Forecast

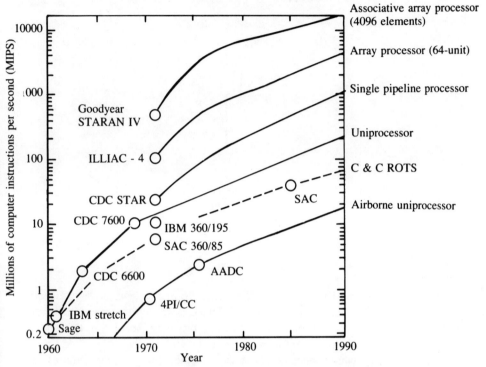

Source: U.S. Air Force, *Information Processing/Data Automation Implication of Air Force Command and Control Requirements in the 1980s* (CCIP-85), Space and Missile Organization (SAMSO), Feb. 1972.

0.0 Clock Time

We can look upon the basic operations in a computer as twofold: *transferring* and *transforming* data. Data are stored in the memory before and after transfer and transformation. The time it takes to move the data in and out of memory can be a significant factor in the cost of operations; therefore, how the data are *organized* becomes almost as crucial an element in the efficient transformation of data as in their expeditious transfer.

No matter how the data are organized, the process of moving the data requires *buses,* which are sets of wires grouped together by similarity of functions and running between functional blocks of a microcomputer system.

Figure 5.15 shows the block diagram of the buses in a microcomputer system. There are three buses: *Data Bus,* 8 bits long, carrying data from the memory to/from the Input/Output (I/O) devices, as well as from the memory to/from

Figure 5.14
MOS Transistor

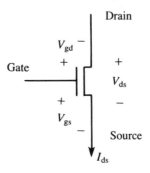

Subscripts in
+ to − direction
sequence.

Source: Carver Mead and Lynn Conway, *Introduction to VLSI Systems* (Reading, Mass.: Addison-Wesley, 1980), 2.

the Central Processing Unit (CPU); *Address Bus,* 16 bits long in the 8080, identifying the particular memory location or I/O device, carrying information from the CPU to the memory or I/O; and *Control Bus,* 6 bits long, specifying the type of current activity, and carrying information from the CPU to the memory or I/O.

The buses are what connect all the devices together. In order to perform complex operations, data have to be transferred and transformed via the buses. The basic system activity is the register-to-register transfer through combinatorial logic, implemented via pass transistors and inverting logic stages.

The system has a basic two-phase, nonoverlapping clock scheme, so that dur-

Figure 5.15
Block Diagram of a Microcomputer System with Buses

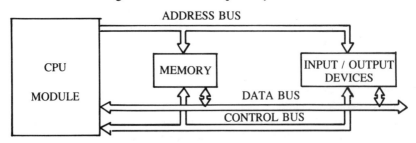

Source: Benjamin C. Kuo, *Digital Control Systems* (Holt-Sanders, 1980).

ing the first clock phase, data pass from one register to a second register; and during the second clock phase, from the second register to a third register, or back to the first register.

What is the minimum clock time required for one such operation? In 1978, it was approximately thirty ns. The target is to reduce it to 2 to 4 ns.

0.0 Length Unit, λ

When conducting materials are laid down on a wafer, such as a silicon wafer, they may not fall precisely where they are aimed, but within a minimum distance of the target. This accidental displacement has to be expected; therefore, the minimum expected accidental displacement when depositing a feature on a wafer is used as the length unit, λ. In 1978, the length unit was approximately 3 microns, and in the 1980s, it is usually about 2 microns (1 micron = 10^{-6} meter). Since light travels about a foot a nanosecond, when the unit is 2 microns, light travels the 2 millimeters of a 1000 λ wire in .006 nanoseconds. The minimum practical value of λ is about 0.1 micron, written as 0.1 μm. In 1981, Lepselter reported a MOSFET with channel length of 0.3 μm and intrinisic switching speed of $\simeq$ 30 ps.[66]

0.0 Switching Energy, E_{sw}

To achieve the best efficiency, maximum speed of accurate signal transmission is often combined with minimum power consumption. Since the maximum power consumption is demanded when the system activity is at its highest, we consider the maximum clock frequency. The switching power per device, E_{sw}, is the power consumed by the device at maximum clock frequency multiplied by device delay. It is the energy dissipated per device per switching event.

0.0 System Parameters

Table 5.4 presents the important system parameters, their 1978 values, and their desired future target values, according to Mead and Conway.

0.0 Balancing of System Parameters

The VLSI technology comprises forming layers of different kinds of materials and depositing collections of materials on them. What are the design goals, and how are they achieved in different architectures?

John E. Price, of Amdahl, presents a comparison of chip architectures against the design goals of: (i) very high speed, (ii) maximum number of pins per chip, (iii) low power, and (iv) short design time. Table 5.5 presents his comparison of five different architectures, designating a match with a design goal as + and a mismatch as − . The gate array logic is the overall winner. As seen from table

66. M. P. Lepselter, *IEEE Spectrum*, Vol. 18, No. 26.

Table 5.4
Important System Parameters

	1978	19XX
Minimum feature size	6 μm	0.3 μm
T	0.3 to 1 ns	≈ 0.02 ns
E_{sw}	$\approx 10^{-12}$ joule	$\approx 2 \times 10^{-16}$ joule
System clock period	$\approx$ 30 to 50 ns	$\approx$ 2 to 4 ns

Source: Carver Mead and Lynn Conway, Introduction to VLSI Systems (Reading, Mass.: Addison-Wesley, 1980), 35.

5.6, the gate array logic dates back to 1965. A sampling of gate array chip architecture by the U.S., West German, and Japanese manufacturers is shown in table 5.7. Since Amdahl is a pioneer in gate array architecture, Price's narration of the revival of the gate array is noteworthy:

> [By the mid– to late '60s] design flexibility at low cost was also provided by means of the master-slice concept,[67] in which a standard array of gates was given a custom logic function by personalizing the metal interconnection between gates.
>
> By the end of the 60s the technology was capable of a level of integration (greater than 100 gates on a chip), at which very few standardized logic functions could be defined. *Gate arrays had fallen into disfavor* (due to lack of adequate design aids for metal routing and testing) limiting custom LSI [Large Scale Integration] to high volume applications. . . .
>
> However, in the early 70s, coincident select techniques and other circuit innovations[68] made possible the realization of large (1 K bits and up) high speed bipolar ROMs [Read Only Memory] which were rapidly put to use in a number of diverse applications, including code conversion, table lookup, arithmetic logic and control logic, (especially in the central processors where ROMs were

67. C. E. Marvin and R. M. Walker, "Customizing by Interconnection," *Electronics*, Feb. 20, 1967, 157–164.
68. C. J. Barrett, et al., "Design Considerations for a High-speed Bipolar Read-Only Memory," *IEEE Journal of Solid-State Circuits* SC-5 No. 5, Oct. 1970, 196–202.

Table 5.5
Comparison of Chip Architecture Against Design Goals

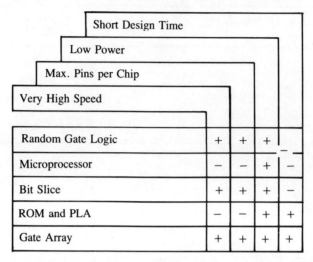

	Very High Speed	Max. Pins per Chip	Low Power	Short Design Time
Random Gate Logic	+	+	+	−
Microprocessor	−	−	+	−
Bit Slice	+	+	+	−
ROM and PLA	−	−	+	+
Gate Array	+	+	+	+

Source: John E. Price, "VLSI Chip Architecture for Large Computers," in *Hardware and Software Concepts in VLSI*, ed. Guy Rabbat (New York: Van Nostrand Reinhold, 1983).

Table 5.6
Chronology of Chip Architecture 1960–1980

	1960	1965	1970	1975	1980
Very high speed			Random gate logic	Gate array Bit slice	
High speed		Random gate logic	ROM Bit slice	Micro-processor PLA	
Medium speed	Random gate logic	Gate array ROM	Micro-processor		

◄— SSI ——— MSI ——— LSI —————►

Source: John E. Price, *op. cit.*, 98.

Table 5.7
Gate Array Product Features

Manufacturer	Gate Delay (nS)	Gate Power (mW)	Gates Per Chip	I/O Pins
Motorola	0.8	4.4	800	60
I.B.M.	0.8	1.7	1500	94
Siemens	0.5	2.0	710	58
Hitachi	0.45	4.0	1500	108
Fairchild	0.4	4.0	2000	160

Source: John E. Price, *op. cit.*, 107.

required to match the logic speeds used in the processor). . . .

For very high speed applications, the *gate array was re-
vived*. A pioneer in this approach was Amdahl Corporation, which used subnanosecond 100-gate array chips in
the central processing unit (CPU) of the 470 V-Series
computer systems.[69]

In the latter part of the 70s, programmable logic arrays
(PLAs) were introduced and used in a variety of applications. PLA and ROM performance ranged from medium
to high speed, but *for very high speed LSI usage was and
is limited largely to gate arrays*.[70] (emphasis added)

0.0 From the Sequential to the Parallel
To appreciate the significance of "subnanosecond 100-gate array chips in the central processing unit (CPU) of the 470 V-Series computer systems," let us look at
the architectural considerations of integrated circuits.

To begin with, the von Neumann machine is one in which the single processor sequentially fetches and executes instructions. By contrast, the Fifth Generation computer is one in which multiprocessors concurrently work, each on an assigned portion of the transfer and transformation of data.

How fast the multiprocessors access the data would depend on how the data
are organized and what process is used to reach the data. The organization of data,
in turn, depends on the understanding of the structure of the data, and the future
demands on the data for decision-making purposes.

69. R. J. Beall, "Packaging for a Super Computer," *Proc. IEEE INTERCON 74*, 1974.
70. John E. Price, "VLSI Chip Architectures for Large Computers," in *Hardware and Software Concepts in VLSI*, ed. Guy Rabbat (New York: Van Nostrand Reinhold, 1983), 97.

Given the organization of the data, there are different ways of splitting them to reduce the delay in accessing them.

The data are accessed so that they may be manipulated by means of logic functions. The basic logic of "And" and "Or" can be implemented via the basic NAND and NOR logic gates. It would be simple to build a large number of such logic gates. However, there will often be encountered important combinatorial logic functions which are expressible in terms of such regular structures.

0.0 Regular Logic Functions and Gate Array

A *gate array* chip is an array of identical and unconnected gates or logic sites in which there are identical arrangements of transistors, diodes, and resistors for forming logic functions. To realize custom logic functions, interconnection patterns are utilized. David Katz of Bell Telephone Laboratories provides a picture of the uncommitted logic array on the left side of figure 5.16; and a picture of the completed logic chip by *custom interconnection* on the right side. In the earlier quotation from Price, the master slice concept was mentioned as one "in which a standard array of gates was given a *custom logic function* by personalizing the metal interconnection between gates," which is shown on the right side of figure 5.16.

0.0 More Complex Logic Functions and PLA

What happens when the logic functions cannot be implemented by means of these regular combinatorial functions? Use Programmable Logic Array (PLA).

In PLA, multiple logic functions are implemented in array form, similar to the permanent placement in memory of the combinational logic functions in Read Only Memory (ROM). Many of the logic functions, however, require only a small fraction of the entire set of outputs stored permanently in ROM, making ROM wasteful. On the other hand, PLA makes it possible to call up only the particular product terms required to implement given logic functions. Logue, of IBM, presents an example of PLA representation, reproduced as figure 5.17. In the third row of the matrix, the product term 3 will be selected when column 5 is a logical zero and column 6 a logical 1. In this case, output columns 13, 15, and 16 will each be logical 1 conditions.

0.0 Significant Savings with LPLA

IBM has found that laser can be successfully utilized in VLSI design. A logical "1" can be personalized on a semiconductor chip by welding two overlapping layers of interconnection metallurgy at the crosspoint with a sequence of laser pulses, as shown in figure 5.18.

How advantageous is Laser Personalizable PLAs (LPLA)?

A comparison of the PLA/laser personalization approach
. . . with an estimate of a conventional approach we have
designed is shown in Figure [5.19]. The result was a *savings of 7.25 man-years and 8 months of processing time,
and a two-pass design* (excluding yield-enhancement passes). The savings reflect the use of PLA macros, the dual

Figure 5.17
Example of PLA Representation

		AND-array inputs	OR-array outputs	
		111	11111	Column numbers
		123456789012	34567	1-17
	1	...|........	|||||	
	2	....|.......	||.|.	
	3	....0|......	|.||.	
	4	.....|......	|..|.	
Product	5	....000|....	.|||.	
terms	6	.....00.|...	.|.|.	
1-11	7	....0.0.0|..	..||.	
	8	..........|.	...|.	
	9	..0.........	|||||	
	10	.0.........0	|||||	
	11	|...........	...|.	

Source: Joseph C. Logue, et al., "Improving VLSI Design Capability," in *Hardware and Software Concepts op. cit.*, 156.

hardware-software modeling approach, and the ability to correct an error on the chip with the laser tool. An LPLA chip for hardware modeling can be personalized rapidly with good yields and meets present and future capacity

Figure 5.18
Connection of Two Metal Layers by Means of a Laser Beam

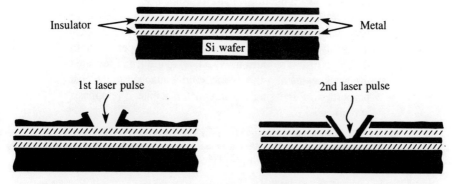

Source: Joseph C. Logue, et al., "Improving VLSI Design Capability," *Hardware and Software Concepts op. cit.*, 164.

requirements. Laser microsurgery is being used to eliminate time-consuming passes through the process line, and
as a diagnostic tool to aid in the rapid isolation and definition of errors.[71] (emphasis added)

0.0 VLSI Technology and Parallel Processing

What does the progress in VLSI technology mean?

The advances in VLSI design and fabrication make it possible to devise
highly parallel processing in which hundreds of thousands of computing elements
can work cooperatively to solve a single given problem. These processors could
be individual chips, several chips connecting hundreds of processors, or even
many processors on one chip, connected by wires. Hayes classifies parallel structures into six: multiple, special functional units, associative processors, array processors, data flow processors, functional programming language processors, and
multiple CPUs.

How do these machine types rate as parallel processors?

Hayes observes that in general, "the price for generality in highly parallel
structures is decreased speed, decreased efficiency of hardware utilization, and increased software requirements."[72]

Mead and Conway point out that our present use of concurrent processing
is limited by the hold that the conventional sequential machine exerts on our
thinking:

> We must approach problems with concurrency in mind,
> recognizing that *communication is expensive and that
> computation is not. . . .*
>
> Perhaps the greatest challenge that VLSI presents to com-
> puter science is that of developing a theory of computa-
> tion that accomodates a more general model of the costs
> involved in computing. Can we find a way to express
> *computations* that is *independent of the relative costs of
> processing and communication,* and then use the proper-
> ties of a piece of hardware to derive the proper program
> or programs? The current VLSI revolution has revealed
> the weaknesses of a theory too solidly attached to the cost
> properties of the single sequential machine.[73] (emphasis
> added)

71. Joseph C. Logue, Walter J. Kleinfelder, Paul Lowy, J. Randal Moulic, and Wei-Wha Wu,
"Improving VLSI Design Capability," in *Hardware and Software Concepts in VLSI*, ed. Guy Rabbat
(New York: Van Nostrand Reinhold, 1983), 173–174.

72. Leonard S. Hayes, Richard L. Lau, Daniel P. Siewiorek, and David W. Mizell, "A Survey of
Highly Parallel Computing," *Computer*, Jan. 1982, 10.

73. Carver Mead and Lynn Conway, *Introduction to VLSI Systems* (Reading, Mass.: Addison-
Wesley, 1980), 330.

Figure 5.19
LSI Productivity for the Eight-bit Microprocessor

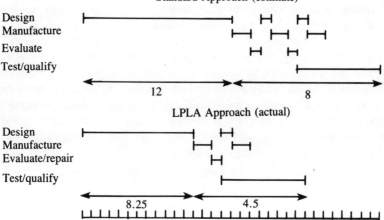

The numbers refer to man-years and each division on the horizontal scale is one month. The net savings were 7.25 man-years and eight months.

Source: Joseph C. Logue, et al., "Improving VLSI Design Capability," in *Hardware and Software Concepts op. cit.*, 174.

While we do not have such a general theory, we would benefit from some considerations that Ullman has put forth regarding parallelism.

Hayes pointed out that decreased speed is part of the price we pay for generality in highly parallel structures. When are we greatly concerned about the speed of the program? Ullman points out that it is when we are most interested in compiling a program into a chip, rather than into machine code for a microprocessor:

> If that is the case, then we would probably be willing to *trade space for time* by putting on the chip more than the minimum number of arithmetic elements and trying to design the chip so that *many operations take place in parallel*.[74] (emphasis added)

A special-purpose chip will run faster than the microprocessor. One reason is that the special-purpose chip saves the time to decode instructions. Much greater speedup is possible by tailoring the data path to the problem on hand, e.g., putting many adders on the chip to perform additions in parallel. Further, each adder can

74. Jeffrey D. Ullman, *Computational Aspects of VLSI* (Rockville, Md.: Computer Science Press, 1984), 376.

be connected by buses to several registers, from which the adder can obtain data and store the results of the transformation.

Ullman discusses "Ideal Parallel Computers and Their Simulation." He notes that in many cases, the model of a computer on which they run is *not* directly physically realizable in the present-day hardware. He finds that three forms of routing are essential to simulate progressively more general "parallel computers": (i) permuatation routing, wherein every processor sends a request to exactly one memory, and each memory has a request, (ii) partial routing, wherein a subset of the processors can each send a request to one memory, but no two processors may request access to the same memory, and (iii) many-one routing, wherein more than one processor may request the same memory, and each processor requests access to some memory. How long does each routing take? O(logn) suggests itself as a yardstick of the time for implementation:

> An interconnection network in which each switch has a fixed number of other switches, memories, or processors to which it is connected requires at least *O(logn)* levels in a computer with n processors and n memories. If not, then some processor cannot even connect to some memory. As a consequence, *O* (logn) time must elapse between the request for data by a processor and the receipt of that data. . . .
>
> Theorem 6.7: There is an interconnection network that allows us to perform any *permutation routing* in $O(log^2 n)$ steps, using only local control. . . .
>
> *[Partial routing]* takes *O(logn)* time with probability approaching 1, where n is the number of nodes in the network. . . .
>
> *[Many-one routing* and getting] the replies to the processors [accomplished] in time *O(logn)* with probability approaching 1. . . .
>
> There has also been recent effort on development of algorithms that run quickly with high probability on parallel computers, such as the butterfly, that have high information transfer rates. . . . There is, in fact, an algorithm due to Ajtai, Kolmos, and Szmeredi[75] for sorting on a parallel machine that uses comparison elements and takes *O(logn)* time in the worst case, not just with high probability.[76] (emphasis added)

75. M. Atjai, J. Kolmos, and E. Szemeredi, "An O(logn) Sorting Network," *Proc. Fifteenth Annual ACM Symposium on the Theory of Computing* (1983), 1–9.

76. Jeffrey D. Ullman, *Computational Aspects of VLSI* (Rockville, Md.: Computer Science Press, 1984), 228, 230, 231, 239, 243.

The trade-off of space for time will clearly benefit from the algorithms that Ullman discusses, and the results that they suggest. Complete one-chip architecture using 1 million transistors per chip is an interim goal of Japan's Fifth Generation Computer Systems, with one-chip architecture using 10 million transistors as the final target.[77]

77. Philip C. Treleaven and Isabel Gouveia Lima, "Japan's Fifth-Generation Computer Systems," *Computer*, Aug. 1982, 86.

PART III
Improving Productivity Through Artificial Intelligence

6

Reasoning Robots for Efficient Enterprises

OVERVIEW: Productivity promises of robotics—the *metal*—depend on the progress in artificial intelligence—the *mental*. Impressive reductions in cost have been made possible by robotics in manufacturing: reduction in workforce by 30 percent at Messerschmidt in West Germany, the handling of 97 percent of the work at Nissan's Zama auto plant in Japan; the reduction by 25 percent of the work-in-process inventory at John Deere Company in Waterloo, Iowa, to mention a few. All of manufacturing, however, represents only 40 percent of the selling price, and direct labor costs only 12 percent of the manufacturing cost, leaving as the target for cost reduction by robotics substitution only 4.8 percent of the selling price. In contrast to this traditional target area for cost reduction through robotics, two other areas together account for 40 percent of the selling price: marketing, sales, and general administrative costs (25 percent), and engineering (15%). To increase productivity in marketing, achievable goals have to be adaptively set (goal-setting), instead of merely meeting goals (goal-meeting). Will artificial intelligence (AI) be able to offer goal-setting capabilities?

The long-term goal of AI is to develop a theory of intelligent information processing. The art of good guessing underlies the successful AI algorithms. The AI methods include: heuristic search, generate-and-test, theorem-proving, backward-chaining, forward-chaining, etc. Imbedding one or more of these methods, many algorithms have been developed to represent human problem-solving behavior. Human problem-solving in chess, logic, and cryptarithmetic is represented in Newell-Simon's Problem Behavior Graph, while computer problem-solving is represented in the Table of Correspondence.

A special kind of problem-solving is robot-planning. The general approach to robot-planning is represented by STRIPS, which models multiple outcomes by associating a probability with each outcome. While in STRIPS the operators state the preconditions, as well as the effect, of application of operators, LAWALY, a LISP program, generates its own procedures to embody the axiomatic description of its capabilities. LAWALY has the robot studying its own environment and ca-

147

pabilities and learning to interact efficiently with it. To both speed up the planning process and to improve the problem-solving capability, PULP-I receives input goal statement in English sentences, instead of the first-order predicate calculus required by STRIPS. PULP-I operates in two modes—learning and planning. After learning from a "teacher" in the planning mode, PULP-I develops a planned sequence of operators to transform the initial model to the desired state. This development of a high-level planning system is a step toward the intelligent control system.

The learning process for robots in all the three instances is iterative, while in humans, the outcome is almost always interactive. The interactive process of problem-solving is found in the Blackboard Model of Expert Systems, which will be discussed.

0.0 From Goal-meeting in Manufacturing to Goal-setting in Marketing

Advances in the metal (robotics) depend on advances in the mental (artificial intelligence). We use the two terms in the generic sense.

In Chapter 4 we discussed several specific advances that have been made by robotics and several specific advances that are possible through robotics. Since the heart of successful automation is the control exercised through communication, we discussed in chapter 5 the many ways in which control is exercised in the metal, including robots. Imbedding controls in the machine, as in machine vision, attempts to make hardware progressively take over software functions; so we investigated the various chip architectures with respect to their promise for highly parallel computing, an essential ingredient of the Fifth-Generation Computers, again using the term generically to mean knowledge-processing systems.

Considerable progress in artificial intelligence (AI) is critical to the creation of the next generation knowledge-processing system. We look at the problems, promises, and prospects of AI with respect to knowledge-processing systems (KPS), from the primary perspective of productivity enhancement.

0.0 Cost Reduction in Manufacturing through Goal-meeting by Robots

The focus of attention in productivity discussions has been on manufacturing. As we saw in chapter 4, manufacturing is responsible for two-thirds of the wealth created annually in the United States. Reducing the cost of manufacturing means that the same amount of wealth is created at less cost. The National Academy of Sciences observes that computer-integrated manufacturing holds the most promise to reduce the cost of manufacturing.

Impressive reductions in cost have been made possible by robotics in

manufacturing. In West Germany the Flexible Manufacturing System (FMS) at Messerschmidt reduced the work force by 30 percent and increased throughput by 25 percent. In Japan, FMS permits robots to handle 97 percent of the work at Nissan's Zama auto plant. Group Technology (GT) saved the John Deere Company of Waterloo, Iowa, $6 million in the first months of operation, while CAD/CAM replaced by 25 percent the work-in-process inventory.

All of these gains have been in the area of manufacturing, as distinguished from merely materials handling. Assembly, for instance, is a manufacturing operation. How much room is there for savings in manufacturing costs? Draper Laboratory studies show that if assembly functions were completely automated, that would only change the total costs by 7 percent for farm machinery, and 4.7 percent for the auto industry. Nonproduction functions, such as engineering, offer far greater potential for cost saving.

The essence of the automated operation in manufacturing is a series of "if-then" actions. The ideal measurements or characteristics have to be met within specified tolerances by each element at the work station before the next operation will be performed on it. If the measurements or characteristics fall outside the allowable tolerances, the element is rejected.

We could look upon the prespecification of the measurement or characteristics as the robot being given a goal to meet, namely, the goal of determining if the presented element meets the goal. The communication of the goal may be by leading the robot through a sequence of steps, or through a sequence of commands in robot control language (RCL), or some similar step. In all cases, some goal is to be met by the robot in the manufacturing process.

Goal-meeting by the robot does result in a reduction of cost.

0.0 Cost-reduction/Profit-increase Potential in Nonmanufacturing Areas

Mike Kutcher, chairman of IBM's Corporate Automation Council, points out that direct labor is usually the target of automation but accounts for only 12 percent of manufacturing cost:

> The enterprise exists principally to develop, manufacture, and market goods for a profit. . . . Consider, for example, the selling price of a typical batch-manufactured product and its several elements of cost [Fig. 6.1]. In this example, manufacturing cost is 40 percent of selling price and the direct "blue-collar " cost is but 12 percent of total manufacturing cost. So even if all the blue-collar tasks were completely automated, less than 5 percent of the selling price would have been affected. . . .
>
> By automating the creation of the physical counterparts— that is, the transform and transfer of parts and mater-

> ials—the enterprise will have automated the blue-collar
> functions (reduced the physical labor content). By auto-
> mating the creation of the various types of data, the enter-
> prise will have impacted the white-collar functions (re-
> duced the mental labor content). By automating both, an
> enterprise can hope to approach the utopian objective of
> efficiently and simultaneously doing different things,
> many of which historically have been in opposition.[1]

If all of manufacturing represents only 40 percent of the selling price, and di-
rect labor costs represent only 12 percent of the manufacturing costs, i.e., $.40 \times .12 = .048$, then robotization of direct labor tasks would only save part of 4.8
percent of the selling price. If robotization could replace all of the indirect labor in
manufacturing, that would represent a saving of part of $.40 \times .26 = .104$, or 10.4
percent of the selling price. Thus, robotization of all direct and indirect labor in
manufacturing can only save part of 15.2 percent of the selling price.

Kutcher shows engineering costs accounting for 15 percent of the selling
price. In chapter 4, we saw that the actual design effort occurs only during 58 percent of the design engineer's day, suggesting that productivity can be increased
significantly by automating all aspects of his work that can be taken over by a
smart machine. The D in CAD will then truly stand for design, instead of standing
for drafting, as is mostly the case now.

The National Academy of Sciences, in its review of integrated computer-
aided manufacturing (ICAM) after five years gave priority 2 to Design/Manufac-
turing Interaction and priority 1 to fabrication. The Academy's schematic of successive stages of coverage of ICAM (see Fig. 4.1) starts with materials handling
inventory, followed by operations before and after materials handling. Double arrows link manufacture and sale, as well as manufacture and purchasing.

The double arrows notwithstanding, the Academy's focus is computer-aided
manufacturing, with the emphasis on manufacturing. Kutcher's diagram allocates
25 percent of the selling price to marketing, sales, and general administrative
costs. Together, engineering and marketing account for 40 percent, making them
equal in importance to manufacturing. It stands to reason that while the total
robotization of direct labor in manufacturing would save a part of the 4.8 percent
of the selling price, engineering and marketing offer a much larger base to produce savings.

0.0 Rethinking "Labor" Productivity

Producing the same goods at lower cost would increase productivity. Output per
hour is the traditional definition of productivity. If 100 units were produced by using 100 units of labor, the productivity index would be 100. However, if direct la-

---

1. Mike Kutcher, "Automating It All," *Spectrum*, May 1983, 41, 43.

Figure 6.1
Elements of Selling Price and Manufacturing Costs

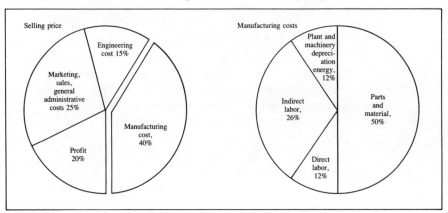

Source: Mike Kutcher, "Automating It All," *Spectrum*, May 1983, 41.

bor, representing 4.8 percent of the selling price were 95 percent replaced by robots, the *labor* costs would be reduced by 4.5 percent. Instead of 100, the labor
input would be 94.44. The productivity index would rise from (100/100 = 1.00
expressed as) 100 to (100/94.44 = 1.059 expressed as) 105.9.

Notice that we only counted labor as the input. Further, the entire cost of
capital in installing the robotic machinery was ignored. It is as though all the direct labor were replaced at zero cost.

If, on the other hand, we had considered the total cost of inputs—machinery,
materials, and manpower, expressed in money terms—the cost of robotization
would be reflected in the computation, and the productivity index would be more
realistic.

Even while we work with "labor" productivity, the impact on productivity of
robotics is understated. Cincinnati Milacron suggests that total productivity of the
piston pin has effectively increased by an order of magnitude when the output per
hour increased from 1400 pins per hour to over 4200 per hour, or the productivity
is 300 percent what it was:

> To begin with, the old line consisted of eight machines
> and required three highly skilled machinists and an in-
> spector. Today the work is done by three machines and
> one operator who needs considerably less technical exper-
> tise.
>
> Today's equipment occupies far less floor space. It is also
> more reliable, averages over 95 percent 'uptime', com-

pared to 70 percent, which means more parts per shift, as well as lower service and maintenance costs. Faster and more reliable production means improved lead times. This, in turn, makes for better control of inventory and more efficient scheduling of assembly.

The quality of the piston pin itself has improved considerably, which means it performs better and lasts longer. Not only is it made of a higher grade of steel, but it's much easier to maintain tolerances for size and roundness. Inspection is automated. There are fewer rejects and less scrap.

Finally the modern line requires less physical and mental effort on the part of the operator. It is quieter, safer, and cleaner. These 'intangibles' can add significantly to worker productivity.

Greater line speed; savings in floor space, manpower, training, and materials; higher quality output; and improved working environment—all of these have a 'multiplier effect' on one another. In the case of the piston pin, it would be safe to say that even though output per hour has increased 'only' 300 percent, total productivity has effectively improved by an order of magnitude.[2]

What Cincinnati Milacron points out is that the manufacturing productivity improvement spills over into related areas, such as materials. Even on the production floor itself, improvements take place, contributing to higher productivity.

0.0 Productivity Potential in Marketing
The raison d'être of production is sale; therefore, the real test of manufacturing productivity is in the market. Does the better quality, lower-priced product attract more customers? Does the increased sales volume increase the net profit and net return on investment (ROI)?

If the answers are yes, the corporation should be able to respond quickly to the increased demand, and capitalize on the advantage offered by automation, including robotics. Even as the new, improved products leave the check-out counter, the sales data should be generating production orders, which, in turn, should be generating materials and parts orders.

There may well be a change in the demand pattern. New customers may buy the product in old locations and/or new locations. Further, old customers may buy more of the new products. Combined with other data, these may indicate the pre-

2. Cincinnati Milacron, *1981 Annual Report*.

paredness of the market to welcome a new version of the present product, and so on.

To perceive these changes in marketing, the incoming data have to be analyzed with respect to the potential products and markets. To miss the nuances of the sales data would be to miss out on the market.

Noting that marketing, sales, and general administrative activities account for 25 percent of the sales price, we recognize how important it is to apply knowledge-processing to that segment of corporate activity. Direct labor accounts for less than 5 percent of the sales price, yet virtually all the attention of productivity promises of robotics is focused on manufacturing. Five times the direct labor share of the sales price is accounted for by marketing.

The requirements are much harder, however, for applying knowledge-processing to marketing. For instance, "if-then" rules would apply to operations such as maintaining inventory (e.g., *if* the stock is less than 30, *then* order 100 units), but to handle a potential demand for a new product in a new market, no simple "if-then" rules would apply. Further, since the outcome would depend not only on the corporation's own action, but equally or more on the actions of other corporations, the rules would have to incorporate probable impact of probable actions on the part of the corporation in anticipation of the probable actions on the part of others.

Can AI assist in handling such situations? Not today. Will AI be able to asist in the future? That depends on the direction of AI, the results, and the application of results. What is needed is knowledge-processing and knowledge-processing systems (KPS). From the success in manufacturing by the goal-meeting capabilities of robotics, we need to turn to goal-setting capabilities in marketing. Here there is greater uncertainty but higher payoffs.

0.0 Agenda for AI—Theory of Intelligent Information Processing

Whether or not AI will develop what we have called goal-setting capabilities, and when AI will develop them, will very much depend on what is perceived to be the long-range view and short-range objectives of the discipline. We will start with the long-range view.

John McCarthy, who coined the term "artificial intelligence" in 1956 as the name for the Dartmouth Conference, the very first conference on AI, said in 1982 that AI is "studying the relationship between problems and the methods for their solution."[3] In our discussion of AI in chapter 8, reference is made to McCarthy's alpha-beta heuristic which makes the finding of a solution to the chess game by following the rule that eliminates from further consideration all moves that one can make which would give one's opponent a strong showing.

3. Philip J. Hilts, *Scientific Temperaments: Three Lives in Contemporary Science* (New York: Simon and Schuster, 1982), 231.

Robotics/Artificial Intelligence/Productivity

The chess game is a problem in which the moves and countermoves are large, but enumerable. What McCarthy does is to eliminate a sizeable segment of the search tree by chopping off branches which represent strong showing by one's opponent. An over-simplification would be to view the alphanumeric as a macro "if-then" rule, *if* offering strong showing to the opponent, *then* eliminate the moves.

Herbert Simon and Allen Newell worked for seventeen years on three denumerable solution-path problems: chess, symbolic logic and cryptarithmetic puzzles. Their 1972 book, *Human Problem Solving*, offers their view of what AI is, and what the objective of their research is:

> This study is concerned with thinking—or that subspecies of it called problem-solving—but it approaches the subject in a definite way. It asserts specifically that thinking can be explained by means of an *information processing theory*. . . .
>
> The natural formalism of the theory is the program, which plays a role directly analogous to systems of differential equations in theories with continuous state spaces (for example classical physics). In information processing systems the state is a collection of *symbolic structures in a memory*, rather than the set of values of position and momentum of a physical system in some coordinate system. Furthermore, a program generally specifies a *discrete change in a single component of the state* (that is, in just one symbol structure) at a moment in time, whereas a differential equation system specifies infinitesimal changes in all coordinates simultaneously. But these are details of the formalism, dictated by the underlying nature of the system under study.[4] (emphasis added)

The Simon-Newell study of problem-solving, and their representation of information-processing by a program, are consistent with their definition of AI: "the part of computer science devoted to getting computers (and other devices) to perform tasks requiring intelligence."[5]

The assertion that thinking can be explained by means of an information-processing theory will probably be widely shared by AI workers. For instance, Edward Feigenbaum says: "What is the most general view of the AI enterprise? Surely, there is the desire to formulate a theory of intelligent information process-

4. Allen Newell and Herbert A. Simon, *Human Problem Solving* (Englewood-Cliffs, N. J.: Prentice-Hall, 1972), 3–5, 11.
    5. *Ibid*, 6.

ing, but there are also other themes. . . . [C]reate intelligent artifacts. . . . codification of knowledge . . . the experiential knowledge of good practice—in short, 'the art of good guessing'."[6]

Intelligence is understandably central to AI, but the orientation is more pragmatic than philosophical. The theory of intelligent information processing that Feigenbaum refers to is for applying it to computer processing, as stated by Simon and Newell: get computers to perform tasks requiring intelligence. Winston would agree: "The central goals of Artificial Intelligence are to make computers more useful and to understand the principles which make intelligence possible."[7]

How would we know when the principles are indeed understood? Alan Turing suggests that the intelligent machine will be able to distinguish between the answers to a set of problems given by humans and machines.[8] The Turing Machine, marked by its knowledge of itself, echoes the hallmark of knowledge identified by Confucius: "When you know a thing, to hold that you know it, and when you do not know a thing, to allow that you do not know it—this is knowledge."

For a computer to understand a message, both the *content* and the *context* are essential. Humans understand data in context. The machine must do this, too, to qualify as being intelligent. Inferences are drawn from the data according to rules given, discovered, and/or modified. Inference is the drawing of all (and only) valid conclusions from the data. How best to organize and retrieve data to facilitate effective inference is the problem of knowledge representation.

0.0 Some AI Problem-solving Methods

McCarthy's alpha-beta heuristic led the way in 1956 with heuristic search as the founding principle of AI. Newell and Simon employ heuristic search in their study of human problem-solving. Several other methods are currently in use in AI. Associated with each method is the organization of data, or knowledge representation, believed to be most appropriate to the inference mechanism employed.

In the **heuristic search method,** an element is selected from the problem space; an operator is selected; a new element of the problem space is attempted, to be produced by the application of the operator to the chosen element in the problem space. The new element is tested to see if it is a solution to the problem. If the test is successful, then the problem is solved. Usually, however, the four processes are repeated: (1) select-element, (2) select-operator, (3) evaluate (new element), and (4) decide-next-step.

The **generate-and-test method** is a method that makes use of the informa-

6. Edward A. Feigenbaum, "The Building Blocks: Artificial Intelligence," *Spectrum*, Nov. 1983, 78.

7. Patrick H. Winston, *Artificial Intelligence* (Reading, Mass.: Addison-Wesley 1977), 1.

8. Alan Turing, "Computing Machinery and Intelligence," Nat. Britain: Manchester University, 1950.

tion available in the problem formulation and tests to see if a potential solution generated is in fact a solution.

There are other methods used in problem-solving, such as the **recognition method,** in which the answer was already in the memory, and was simply recognized as soon as the question was posed.

The **theorem-proving method** proves (or disproves) a main theorem, using a logical calculus in terms of which the goal is expressed. When humans prove theorems, they use specialized knowledge to select previous proofs useful to the present theorem-proving. They also break down the given theorem into subproblems, the solutions to which would move them closer to the solution of the main theorem.

The **backward (forward)-chaining method** connects the goal back to theorems already proven, or connects the known conditions forward to the goal, while the **means-ends analysis method** works both forwards and backwards.

0.0 Some AI Algorithms

Central to these, and other AI methods, is the notion of learning. Can the goal be reached faster, in fewer steps? Can the iterations be adaptive—applying the knowledge gained about the relationship between the inference and the goal to progressively improve the inference process? Let us review some of the AI algorithms that have been developed.

0.0 AI Algorithms: (1) Problem Behavior Graph (PBG)

Whether the human problem-solving is viewed in terms of the heuristic method or one of the other methods, or some combination, the problem-solving *behavior* can be represented in terms of the direction in which he moves with respect to his state of knowledge. The advantage of the representation of the Problem Behavior Graph is that the application of an operator to the state of knowledge can be identified whenever it occurs. Time is on the x-axis, making the graph linearly ordered by the time of generation. The Problem Behavior Graph is reproduced as figure 6.2. The problem solver is always located at some node in the figure. Since each node represents a state of knowledge, the problem solver is supposed to possess the information about the problem as defined by the node. By applying a method to an element of the problem space, he acquires new information. According to Newell and Simon:

> The subject starts at node 1 in the upper left-hand corner.
> The first operator applied is Q1, producing the state of
> knowledge indicated by node 2. Then Q2 is applied,
> leading to node 3. At this point the subject returns to the
> same state of knowledge as in node 2; this is shown in
> node 4. Since the act of returning did not involve an

Figure 6.2
Problem Behavior Graph

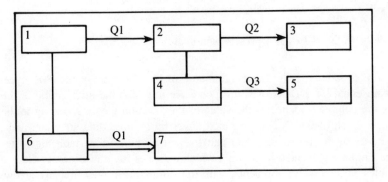

Source: Allen Newell and Herbert A. Simon, *Human Problem Solving* (Englewood Cliffs, N.J.: Prentice Hall, 1972, 1973.

operation—such as recalling the prior state, or abandoning the information produced by Q2—the move from node 3 to node 4 is not shown as an operation. At node 4, Q3 is applied to produce node 5, and then the subject returns (node 6) to the state of knowledge represented by node 1. Node 6 must be written on the line below nodes 4 and 5, since it occurred later in time. Q1 was applied again, as indicated by the double line emanating from node 6. The identity of the knowledge states at node 7, 2, and 4 is not indicated in the graph.[9] (See figure 6.2)

A *graph* (network) comprises nodes, some pairs of which are joined by branches (links, paths). When the direction of the branches is indicated, it is a directed graph (digraph). When there is activity (movement) along the branches, that constitutes network flow.

0.0 AI Algorithms: (2) Table of Correspondence

Not only human problem-solving but also computer problem-solving can be represented by networks. The Table of Correspondence specifies the particular value of the function $y = y(k)$ that will be assumed by y at point k. It can also be represented as a *graph*, when the successive values of the variable are specified in terms of the rules of transition.

How can the transition of x from one point in time t-1 to the next, t, be mathematically specified?

9. Newell and Simon, *op. cit.*, 173.

If the state of the system at each discrete moment is uniquely defined by its state in the preceding moment and by its input at the preceding or current moment, we have P-P (past-past) automaton:

$$x^{p+1} = F(x^p, P^p), p = 0, 1, 2, \ldots$$

Or, we have the P-Pr (past-present) automaton:

$$x^{p+1} = F(x^p, P^{p+1}), p = 0, 1, 2$$

The equations to the finite automaton of the P-P type or the P-Pr type may be represented by a Table of Correspondence in which the intersection of the row x and the column y identifies the value of the function y corresponding to the values of x, each of which is determined either by its preceding state and preceding output, or by its preceding state and *present* output.[10] This table of correspondence can be represented by a graph which indicates the transition of values of the variables in successive moments. The transitions are represented by arrows between circles representing the values of the state of the variable x. Only one arrow can *start* at each circle, but any number of arrows may terminate at it.

0.0 AI Algorithms: (3) Automation of Organization of Algorithmic Languages

Intelligent machines infer by applying rules of inference to data. Since the machines have to "understand" the rules, they must "understand" some language, robot control language (RCL) being an illustration.

To the extent that the programs can be written in languages that represent the mathematical and logical processes of inference, to that extent the use of the computer capability would correspond directly to the electromechanical composition of the computers themselves. Progress, however, in the employment of automata to aid human problem-solving depends on the success of the formal language in describing the concepts and processes.

The need to organize programs in the computer in such a way that "housekeeping" tasks are handled with a minimum of human intervention was recognized in the development of List Processing (LISP) languages. They assist the programmer in two major ways: (1) organizing the computer memory as an *associative store*, which permits flexible cross-referencing, or associating one list with another; (2) providing *closed subroutines*, to build from many subroutines a many-level hierarchical program structure.

Academician Victor M. Glushkov points out that the formal algorithmic languages, such as LISP, ALGOL, COBOL, are each oriented to different classes of applied problems. He visualizes the development of the algebra of algorithmic languages to reach a state in which it passes into common knowledge, instead of being the private preserve of programmers. Such a development would permit the equivalent transformations of expressions in those languages. While the develop-

---

10. Mark A. Aiserman, Leonid A. Gusev, Levi I. Rozonoer, Irina M. Smirnova, and Aleksey A. Tal, *Logic, Automata, and Algorithms* (New York: Academic Press, 1971), 62–63.

ment of that algebra may be a generation away, the Institute of Cybernetics at the Academy of Sciences of the Ukranian SSR works on representing an arbitrary algorithm in the form of a scheme of interaction of two automata:

> The basic theorem in the theory consists of the possibility of reducing to regular form an arbitrary algorithm program that uses the same system of elementary operators and elementary conditions. When written in such a form, the algorithm becomes an element of algorithms $\mathcal{K}$ expressed in terms of the generating elements of that algebra. When we have written the system of defining relationships of the algebra $\mathcal{K}$, we can carry out transformations of regular programs of that algorithm, finding the programs that are best from some particular point of view.[11]

Glushkov feels that it is now possible to automate the organization of algorithms, instead of performing random empirical searches. The organization is not completely automatic; it requires man-machine interaction in which the man chooses the better among the search schemes.

0.0 AI Applications to Robotics

We said earlier that the payoff from knowledge-processing in areas other than manufacturing may be greater, since manufacturing represents 40 percent of the selling price, while marketing and engineering together also account for 40 percent of the selling price. Most effort to date, however, has been aimed at reducing a portion of the manufacturing cost, i.e., direct labor, which accounts for only 4.8 percent of the selling price.

Shades of AI have already appeared in our discussion in chapter 4 of robots, CAD/CAM, and other automation technologies in integrated manufacturing. Further, in chapter 5, the theme of implementing controls in automation owed much to AI.

We will focus on robot planning, as a special kind of problem-solving, for which the problem-solving methods and AI algorithms discussed in this chapter have special significance. We used the term *goal-meeting* to designate the robot matching observed characteristics, features, and/or measurements with prespecified characteristics, and on the basis of the match, accepting or rejecting the observed item. Goal-meeting in the case of machine vision is multidimensional. Visual inspection can be viewed as pattern-recognition, the methods for which form three groups: (i) decision-theoretic or discriminant, (ii) template-matching, and (iii) syntactic and structural approaches. Powerful shades of AI abound indeed.

---

11. Victor M. Glushkov, "Problems in the Theory of Automata and Artificial Intelligence," *Journal of Cybernetics*, Vol. 1, Jan.–March 1971, 103–4.

What AI applications to robotics are our present focus?

Planning, in general, and high-level planning in particular. When the goal is specified, planning charts a sequence of actions to meet the goal. If the goal is not specified, however, but needs to be formulated, that would require goal-making, not merely goal-meeting. The shift toward the former would offer great potentials for cost reduction and profit increase in the marketing and engineering components of corporate activity.

0.0 General Approach to Robot Planning: STRIPS

In 1971 Fikes, Hart and Nilsson presented the Stanford Research Institute Problem Solver (STRIPS), which relaxed many of the strict ground rules dealing with only a single robot working on a single problem.[12]

A set of statements describing the initial environment, e.g., robot is in room R3, one of the three rooms connected by doorways, and a set of operators for manipulating the environment, e.g., PUSHRM—BX, RX, DX, RY—which pushes object BX from RX into adjacent room RY through connecting doorway DX, define the problem environment.

The problem is to achieve a goal. To achieve the goal, STRIPS has to find a sequence of operators which transform the initial model, or state, into a final state in which the goal is provable.

The STRIPS operators state the preconditions under which each operator may be applied (e.g., robot must be in room R3 before it can move BX), and the effects of the operator. It will be illegal to push BX from R3 if the robot is not in R3. Knowing why it is illegal enables STRIPS to apply the operator GOTORM (R3). The proof cannot be completed when the operation is illegal, but instead of discontinuing the search for a plan to meet the goal, STRIPS extracts a 'difference' between the initial model and the goal indicating a set of statements that would help complete the proof, such as GOTORM (R3).

Multiple Outcome Modeling is possible with STRIPS by associating a probability with each outcome. Each node of the search tree represents a state, and each branch from a node represents an outcome of an operator applied to the node.

The goal is stated as a well-formed formula (WFF). Multiple goals are defined by a set of WFFs, each possessing an "urgency value" measuring its relative importance. The overall objective of the system is to maximize some sort of benefit/cost ratio.

When should the planning program stop? When the goals are achieved, i.e., when the theorem is proven, or when the theorem cannot be proven. In the case of multiple objectives, when the search tree has been expanded sufficiently to determine that no plan in the tree will be able to prove the theorem, it is an appropriate stopping point, although absolute failure is not established.

12. R. E. Fikes, P. E. Hart, and N. J. Nilsson, "Some New Directions in Robot Problem-Solving," *Machine Intelligence*, Vol. 3, 1972, 405–30.

Figure 6.3 presents an example of multiple-outcome. The robot should move box B1 into room R1; there should be no box in the same room with wedge W1. The initial condition states that the robot does not know if W1 is in R1, and if doors D1, D2, and D3 are open or closed. The operator CHECKDOOR (DX) applied to D3 in figure 6.3 may find it open or closed. If it is closed, then the operator is applied to D2, and so on. When should the search stop? If the theorem is proven, disproven, or when all unexpanded nodes have probability less than say, .2, as in figure 6.3.

How can the planning be generalized?

Fikes, Hart, and Nilsson presented a novel format to store plans which have general application.[13] The method is to replace specific *constants* in the solution by *parameters*, so that the plan found by STRIPS will be applicable to a whole *family of tasks*. The execution monitor, PLANEX, uses the generalized macro-action, called MACROP, to guide execution.

0.0 Procedural Approach to Robot Learning: LAWALY

In STRIPS the operators state the preconditions and the effect of application of operators. LAWALY, a LISP program, generates its own procedures to embody the axiomatic description of its capabilities. These procedures work best if the tasks have a hierarchical structure.

Just as in STRIPS, the environment is specified in a set of predicates, such as AT (BOX 1, A1). These initial states can be changed to other states by applying operator(s), if the preconditions are satisfied for the application. LAWALY builds a procedure for each operator and links it with the system monitor. The preconditions that the operator must satisfy are reordered according to the hierarchical order of the subtasks. If several subtasks are to be accomplished, some have to be accomplished before the other(s). LAWALY will try to solve the highest-ranked tasks first, moving to the successive lower tiers of the hierarchy.

LAWALY would probably improve its speed by building macrooperators. Even without macrooperators, LAWALY found that the STRIPS world which had no task requiring more than fifteen steps was no challenge. As Table 6.1 shows, LAWALY took much shorter time to perform the same theorem-proving as STRIPS, and took on tasks *d,e*, and *f* in addition to the tasks done by STRIPS, viz., *a*, *b*, and *c* in table 6.1.

The concept that LAWALY offers is an approach to robot learning:

> Our approach to robot planning can be viewed as a form
> of learning: the robot studies her own environment and
> capabilities, and learns to interact efficiently with it. We
> can call this type of learning *procedural learning*, and
> contrast it with statistical learning—and structural learn-

13. Richard E. Fikes, Peter E. Hart, and Nils J. Nilsson, "Learning and Executing Generalized Robot Plans," *Artificial Intelligence*, Vol. 3, 1972, 252–88.

Figure 6.3
Multiple Outcome Search Plan

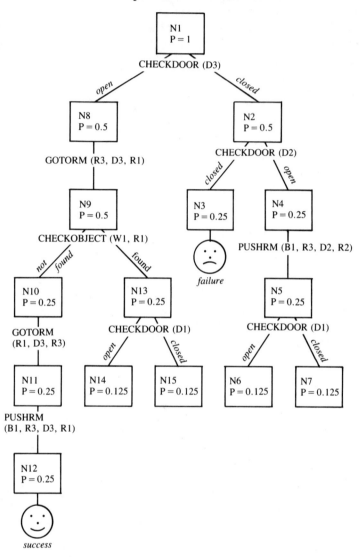

Source: R. E. Fikes, P. E. Hart, and N. J. Nilsson, "Some New Directions in Robot Problem Solving," *Machine Intelligence*, Vol. 3, 1972, 425.

Table 6.1
Task Performance Comparison: Strips vs. Lawaly

Task	Theorem proving time in seconds		Number of operator applications				Time per node in the search tree in seconds		Number of nodes in LAWALY's initial world
			On solution path		the search tree				
	STRIPS	LAWALY	STRIPS	LAWALY	STRIPS	LAWALY	STRIPS	LAWALY	
a Turn on the lightswitch	46.5	1.627	6	4	6	4	7.75	0.407	24
b Push the three boxes together	92.5	4.095	4	4	6	5	15.42	0.819	24
c Go to a location in farthest room	103.0	2.625	5	5	5	6	20.6	0.438	24
Some other problems given to LAWALY in the same world.									
d Push the three boxes together symmetrically (Impossible task in model)	—	47.796	—	no soln. found	—	50	—	0.956	24
e Go next to box 1 and box 2 without pushing them together (physically impossible)	—	3.114	—	3	—	3	—	1.038	24
f Turn on the lightswitch, push box 1 and box 2 next to the lightswitch and go to a location in another room	—	10.268	—	15	—	16	—	0.642	24

Source: L. Siklossy and J. Dreussi, "An Efficient Robot Planner Which Generates Its Own Procedures," *Proc. Third Int. Jt. Con. on Artificial Intelligence,* William Kaufman & Co., 1973, 430.

ing—where improvement in performance results from the building and modification of structures. Many works in pattern recognition and checker-playing program of Samuel are examples of statistical learning. The generation of MACROPs may be considered a form of structural learning; other examples are [in 14 and 15].[16]

0.0 Supervised Learning for High-level Robot Planning: PULP-I

The core of robot learning in STRIPS is the substitution of constants by parameters to develop MACROP, making the particular plan formulated by STRIPS applicable to a whole family of tasks.

The plan formulation in STRIPS is the use of the resolution theorem prover, which may generate irrelevant and redundant clauses, which, in turn, would lengthen the search for an appropriate sequence of operators.

Both to speed up the planning process and to improve the problem-solving capability, PULP-I receives input goal statement in English sentences, instead of first-order predicate calculus required by STRIPS.

PULP-I operates in two modes—learning and planning. In the learning mode, a teacher provides input knowledge, C_e, which is transformed into C_i, collection of task knowledge in internal or semantic representation. That transformation includes the analysis of the procedural knowledge and its storage in the system memory. The system terminates the learning mode with the response "DONE."

When a command sentence is put into the system, PULP-I operates in the planning mode, which forms a plan that transforms the present state into the specified state. When the sequence of operators accomplishes the state transformation as specified, the different candidate sequences are compared in terms of their semantic matching. The one with the smallest value is tried first to develop a solution plan. If it fails, the next level sequence is tried, until a solution plan is found, or exhaustion of choices is reported.

What is the advantage of PULP-I?

> The system dubbed PULP-I is able to accumulate knowledge through the process of learning and to conceive a plan consisting of a proper sequence of operators in a space of world models to transform a given initial model in which a given command is satisfied. . . . The learning

14. L. Siklossy, "Natural Language Learning by Computer," in *Representation and Meaning: Experiments with Information Processing Systems*, ed. H. A. Simon and L. Siklossy (Englewood Cliffs, N.J.: Prentice-Hall, 1972).

15. P. H. Winston, "Learning Structural Descriptions from Examples," MAC Report TR-76, M.I.T., 1970.

16. L. Siklossy and J. Dreussi, "An Efficient Robot Planner Which Generates Its Own Procedures," *Proc. Third Int. Jt. Conf. Artificial Intelligence*, William Kaufman and Co. 1973, 426.

system is an intermediate one between rote learning and generalization learning, and is based on the concept of analogy. . . . This development of a high-level planning system is a step toward the intelligent control system originally proposed by Fu [in 17,18].[19]

0.0 Back to the Blackboard!

Fu's reference to an "intelligent control system" brings us back to chapter 5, which opened with the observation that at the heart of successful automation of manufacturing is the control exercised through communication. We have elsewhere defined control as "the influencing of the behavior (motion) of an entity in a prespecified manner by means of elements within or without the entity."[20] The prespecified behavior of STRIPS, LAWLEY, PULP-I, and other robot problem solving systems is twofold: (i) correct and efficient solution of the particular problem, (ii) operational generalization of the solution process to a family of similar problems.

The learning process in the three instances are iterative. In humans, though, the outcomes are almost always interactive, since no one participant controls all the variables affecting the outcome: e.g., even if one were the sole producer of a good, the demand for the good rests with outside elements.

Recent experience with Expert Systems lends credence to the importance of the interactive process of learning. As Feigenbaum puts it:

> In this author's view, the most powerful of these control structures for search is the blackboard model. In this scheme multiple sources of knowledge cooperate by means of a common knowledge structure that represents the emerging pieces of the solution. Each body of knowledge looks at the "blackboard" to see what it can do to build onto the emerging solution. The system achieves power from its capability to mix arbitrarily forward and backward chaining strategies, using bodies of knowledge represented at different levels of abstraction.[21]

In the next chapter, we will look at how, with promising results, the blackboard model applies to different expert systems.

17. K. S. Fu, "Learning Control Systems—Review and Outlook," *IEEE Trans. Autom. Control*, AC-15, April 1970, 210–21.

18. K. S. Fu, "Learning Control Systems and Intelligent Control Systems: An Intersection of Artificial Intelligence and Automatic Control," *IEEE Trans. Autom. Control* AC-16, Feb. 1971, 70–72.

19. Supachai Tangwongsan and K. S. Fu, "An Application of Learning to Robotic Planning," *Int. Jo. Computer and Info. Sciences*, Vol. 8, No. 4, 1979, 333.

20. George K. Chacko, *Computer-aided Decision-Making*, Elsevier, NY, 1972, 77.

21. Edward A. Feigenbaum, "The Building Blocks: Artificial Intelligence," *Spectrum*, Nov. 1983, 78.

7

Knowledge-Processing for Better Decision-making

OVERVIEW: The 1983 program of the Defense Advanced Research Projects Agency (DARPA), called Strategic Computing and Survivability, is committing close to $1 billion to develop by 1989 machines with humanlike capabilities, allowing them to see, reason, plan, and even supervise the actions of military systems in the field. Why is DARPA committing megabucks to artificial intelligence? Because of the track record of expert systems in giving high-level advice or carrying out a complex function normally associated with human expertise.

Planning, human-style, is universally applicable to both profit-making and nonprofit-making situations, although it is easier to develop the concept with reference to the former. Planning balances the demand and supply—the external demand for fulfillment characteristics of the goods and services by the internal technological performance capabilities of the corporation. Two out of the four "humanlike capabilities" envisaged for the 1989 machines are covered by planning and its implementation—supervision. A third capability, seeing, enables the robot to interact with external environment; in particular, it is able to exercise control by acting on the visual information.

"Reasoning" is the underlying capability which enables the exercise of sight, planning, and supervision, leading to the definition of reasoning: the process of abstraction and/or application of abstraction to entities sharing similar characteristics for identification and/or action. The twin objectives of reasoning—identification and action—underscore basic differences in the outcome of the reasoning. The solution in solution-oriented reasoning is a unique symbol fulfilling the specified conditions and characteristics; the solution in structure-oriented reasoning is the optimal sequence of current resource commitment for subsequent results. Interacting variables and interacting participants in structure-oriented reasoning problems demand far higher degrees of sophistication than those required for solution-oriented reasoning situations. Expert systems can handle both types of problems.

Granted that expert systems embody human expertise, what is expertise?

Again, granted that the output of expert systems should be useful to the user, what determines usefulness? Errors are inevitable when resources are committed, knowing that one of a number of outcomes could occur, but not which one. What is most useful to the user is the expert system which will aid him in minimizing the graver consequences of wrong action. The concept is illustrated with respect to the most celebrated expert system, PROSPECTOR, which identified $100 million worth of molybdenum deposit, and CADUCEUS, which accurately diagnoses diseases from symptoms. A new definition of expert systems is developed, and major expert systems are discussed. Almost all known expert systems employ iterative processes, while what is critically needed for important real-life problems affecting productivity is interactive problem-solving capability. The emerging subdiscipline of Distributed Problem-solving seeks to develop the welcome capability of processing incomplete and erroneous input data at each autonomous node in cooperation with each other to arrive at solutions to a single problem.

0.0 The Expert and the Intelligent

Planning, which we identified in the last chapter as an important aspect of robotics research, is one of the four humanlike capabilities of the "new generation of superintelligent computers for military use." The 1983 program of the Defense Advanced Research Projects Agency (DARPA) called Strategic Computing and Survivability plans to develop this capability:

> These machines will have humanlike capabilities, allowing them to see, reason, plan, and even supervise the actions of military systems in the field. High-performance VLSI-based hardware and multiprocessor software will be developed with the goal of demonstrating them in selected military applications in the field by the end of the 1980s. Close to $1 billion may be spent over the rest of the decade toward this end.[1]

0.0 A Billion-Dollar Commitment

A billion dollar investment in research in seven years is indeed impressive. Advances in artificial intelligence are central to the realization of computers with "humanlike capabilities." And DARPA has almost single-handedly supported major AI efforts for decades, being responsible for the development of the world's first supercomputer, ILLIAC-IV in the early 1970s, pioneering of timesharing and packetswitching, especially via ARPANET, initiating VLSI design innovation and fabrication training, to mention a few.

1. Robert S. Cooper and Robert E. Kahn, "SCS: Toward Supersmart Computer for the Military," *Spectrum*, Nov. 1983, 53.

Why is DARPA committing megabucks to AI?

Former deputy director of DARPA, now assistant secretary of technology for the Air Force, Dr. Robert E. Kahn was concurrently the director of the Information Processing Techniques Office from 1979 to 1984. He points out that AI has only recently developed "a sufficient track record of accomplishment to attract industrial interest":

> The commercial sector, the military, and major sectors of the health and education fields are now exploring techniques developed by AI researchers. The most notable example of accomplishment is in developing expert systems to give high-level advice or to carry out a complex function normally associated with human expertise.
>
> Examples of such systems are most prevalent in the medical area, but significant investments have also been made in such areas as equipment configuration and signal interpretation. Important earlier work in this area led to expert systems in molecular spectroscopy and symbolic mathematics. These systems incorporate numerous facts and heuristics (rules of thumb) about their domain of expertise and arrive at their conclusions through a kind of systematic application of these facts and heuristics to narrow the possibilities to explore. In this way it is often possible to arrive at a specific expert conclusion "heuristically," even with a small set of rules of thumb.[2]

Kahn goes on to point out that the expert systems are restricted to simple problems. Nevertheless, it seems that expert systems have proven themselves to be practical.

0.0 Intelligent? Not Really

We will be discussing at some length what the expert systems are, what they can do, and what they cannot do. First, the near-term research targets of DARPA merit another look. Seeing, reasoning, planning, and supervising are capabilities not possessed by the expert systems today. What does the future hold? Will the machines be able to reason?

In chapter 6, we identified the agenda for AI as intelligent information processing. Expert systems can narrow the amount of information processing, but are they intelligent?

It is interesting to note what Kahn visualizes for the future:

---

2. Robert E. Kahn, "A New Generation in Computing," *Spectrum*, Nov. 1983, 38.

> Intelligent systems will begin to make their way into the world, but few people will consider them to be really intelligent after all. The fundamental problems of deep reasoning and understanding will become defined but will elude all attempts to solve them. Expert systems of many kinds will become available, and limited forms of computer expertise will begin to be sold in the marketplace. . . .
>
> Completely new applications will emerge from this wealth of new technology, and computers will begin to play more important roles in planning and management throughout society. Deep strategic or tactical insights by machine may still be decades away, but the competitive use of such planning and management systems will foster even more widespread use of them, along with demands for their continual improvement.[3]

The "deep reasoning and understanding" that Kahn refers to probably best separate the man from the machine. Even without humanlike reasoning and understanding, expert systems will enable computers to "play more important roles in planning and management throughout society." There is that word again: planning. "Reason, plan and even supervise" aptly describe critical aspects of "management."

0.0 Planning, Human-Style

The important roles for the computer "in planning and management throughout society" that apparently undergird the billion-dollar DARPA commitment of AI research in the next seven years make it imperative to clarify what planning is, and what it is not. The discussion should preferably tie planning to management.

While planning, like management, is universally applicable to profit and non-profit institutions alike, it is easier to discuss the concept in terms of profit-making business.

Profit makes business go. Without profit, there would be no production and no distribution in a market economy. Production is for sale. There is no sale without demand for the products and services. The moving force in a market-economy, therefore, is demand.

The demand is external to the corporation. The world outside the corporation makes the demand for the products and services of the corporation, which can be described as the demand to fulfill the needs and wants. Whether Pepsi-Cola can fulfill the demand for thirst-quenching with taste as well as, if not better than, Coca-Cola is something both Pepsi and Coke constantly and vigorously fight

3. *Ibid.*, 40, 41.

170

about, because it is on their success-to-persuade that they offer the specific *fulfill-ment characteristics* on which their corporate success, and indeed their very survival, depend. Whether it is a product (goods) or a service, the external demand is for the fulfillment characteristics of the goods and services, as perceived, correctly or incorrectly, by the consumer.

The perception by the consumer is fundamental to the demand. It is not the chemical composition of the cola printed in small type on the bottle that spells fulfillment, but the jingle blitzed over the media: "Coke is the real thing," whatever that may mean to the consumer. The medium is the message; and the message is the promise of fulfillment. As long as the performance is within the threshold of the promise to each consumer, the demand is sustained.

Competition is not confined to external competitors. Perhaps Kool was one of the earliest examples of competition from *within*, one brand of cigarette by the same manufacturer competing head-on with another brand by the same company, not only in the market, but also in the advertising. The '50s advertisement of Kool pitting it against its sister cigarettes from the same company underscored the importance of *perceived* fulfillment characteristics.

As tastes change, advertisements change. A figure-conscious public is promised the cola taste without the calories—"fills you up, not out." Two or more products from the same company simultaneously offer the real thing and the fake thing and succeed in both equally well.

The corporation must have the *technological capabilities* to meet the demand for the fulfillment characteristics, in this instance, to produce the real thing and the fake thing. Having created a stable niche in the market with the real thing, the corporation, sighting the rising thirst for the hour-glass figure, can go out and acquire the technological capabilities to meet the fulfillment characteristics of the fake thing.

The technological capabilities comprise all the capabilities that make and meet the demand, from the raw material and the patented formula to the world-wide bottling plants and the world-wide jingles.

Planning balances the demand and supply—the external demand for fulfillment characteristics of the goods and services by the internal technological performance capabilities of the corporation.

The demand is the demand for tomorrow. It is the *expected* demand that planning is concerned with. Today is too late for planning. And if planning is not done by design, it is done by default. Default occurs when the expected demand is not determined adequately. Even with accurate determination of tomorrow's demand based on yesterday's and today's data, change in taste tomorrow may well upset the preparation to meet tomorrow's demand. Planning is no guarantee of success. Without it, though, the chances of making and meeting market demand are slim; with it, the chances of learning how to make and meet market demand are great.

We can now define the activity that is at the heart of management. While it is

goods and services that the corporation produces, long-term corporate survival depends on creating *value*. Value is extrinsic. The 50¢ that a customer pays for a can of coke is the measure of monetary value, the vote with the pocketbook. The customer creates value, *not* the corporation. All the technological performance capabilities in the world will not create a cent of value; it has to come from outside the corporation. The 50¢ could have been paid to buy a Coke, a Pepsi, or a host of other liquid refreshments. When the can chosen is Coke, its producer gets the value. Included in the 50¢ is the reimbursement to the corporation for all the costs of technological performance capabilities plus profit; and profit makes business go.

To survive and succeed, management has to commit corporate resources today for market results tomorrow. Products come from within; profits come from without. The management controls production variables; the market controls profit variables. But neither controls in full either set of variables, so profit-making is an interactive process. Production for profit is planned by management and disposed by the market. Planning is the battle plan to acquire territory through technology. It sets the objective in operational terms. If the objective is achieved, the successful strategy can be applied again. If the objective is not achieved, learning why enables the development of a successful strategy.

We can define planning, then, as: *The process of allocating and/or acquiring present and/or potential technological performance capabilities to meet and/or exceed expected external demand for fulfillment characteristics.*

0.0 Reasoning, Human-Style

How will we know it when AI has developed "humanlike capabilities"?

The four capabilities specifically identified for the near-term—by 1989—are seeing, reasoning, planning, and supervising." Under our definition of planning, supervision becomes adaptive implementation of the battle plan, be it in the military or in the market battlefield. In chapter 5, we discussed the capability of *seeing* identifying the state-of-the-art of machine vision that enables the robot to interact with external devices and sensors. We discussed the capability in terms of current developments in the five principal areas of *sensing, segmentation, description, recognition*, and *interpretation*. At an IEEE workshop, Industrial Applications of Machine Vision, held in May, 1982, John F. Jarvis, head of the Robotics Systems Research Department at Bell Laboratories, summarized the areas of needed research. Jarvis also defined robot vision as a means of dealing with the *unexpected*. Insofar as our definition of planning states that it is a process dealing with *expected* external demand, the planning capability is the complement of the seeing capability.

Reasoning is the capability of coping with both the expected and the unexpected. It is exercised through abstraction and application, and it can be either inductive or deductive. Reasoning can be descriptive, prescriptive, or associative. Reasoning frequently employs analogy to apply what is already known to the

newly encountered entity (object, event, behavior, relationship) or to determine its inapplicability, as is done in the human act of seeing.

The act of seeing does not take place in the eye; it takes place in the brain. The visual faculties register the stimulus, but the brain recognizes it as a previously encountered entity, as something not quite similar, or as something not classifiable with the available record of entities. As we pointed out in chapter 6, vision in general, and visual inspection in particular, can be viewed as pattern-recognition, the methods for which are directly applicable to the objectives of planning and supervision. Planning specifies what ought to be; supervision specifies what is; and control specifies how the latter can be changed into the former through the allocation of resources to alter results. Management exercises control to move the corporation ever closer to what-can-be, as specified by planning.

The four capabilities specifically identified for the near-term act together to achieve control. Reasoning is the underlying capability which enables the exercise of the other three. We can define reasoning: *The process of abstraction of, and/or application of abstraction to, characteristics of entities for identification and/or action.*

Identification specifies *what is*, e.g., the soft, round object is an orange; the three bars are from Beethoven's Fifth Symphony; and disease is emphysema. Action specifies *what does*, e.g., to capture military objective 1, follow plan A, failing which, shift to plan B; to capture market objective 1, follow plan A, failing which, shift to plan B.

0.0 Iterative and Interactive Approaches to Identification and Action

The twin objectives of reasoning—identification and action—underscore basic differences in the outcome of the reasoning process.

0.0 Solution-oriented Reasoning

The outcome of reasoning in the case of identification is unique, or near-unique. There is only one answer possible. That answer is the conclusion arrived at from the given premise(s) and evidence. If the question is mathematical, the solution is unique-valued: there is only one value which satisfies the equation.

What is the objective of the *user* of the reasoning?

If he (she) wants to specify what the entity is, his objective is satisfied when the single value is found. The three bars of music are in Beethoven's Fifth Symphony; they are from nowhere else.

In such a problem, the outcome is fixed. The chemical compounds do not change their structure while their identity is reasoned from the mass spectrographic data; the pulmonary disease does not change its nature (though magnitude may change over time) while its identity is reasoned from laboratory data.

The argument also applies if the outcome is not a single, unique value, but

say, a small number of values, such as two or three. In the solution to the quadratic equation, the unknown variable can take the same value in the plus or minus direction, both of which are admissible. The identified molecular structure could apply to more than one compound. The pulmonary disease could be one of two or three varieties.

The fixed outcome of the reasoning makes it possible to arrive at the outcome by *iteration*. Given that the destination is fixed, it can be reached by repeated process of zeroing in on the target, such as the circling in a helicopter over the area of the destination. By observing the distance from the destination in each pass, the sweep of the helicopter can be made smaller and smaller, until pin-point landing is made. In this case, we know the destination and can identify it (e.g., a farmhouse) and circle closer and closer to it.

The one assurance needed is that the outcome is fixed. *If* the farmhouse is assured to be in the area circled by the helicopter, or if the farmhouse is one of the three structures in the area, the outcome can be considered fixed for purposes of searching for it and reaching it.

The variables involved in the search do not have to be limited to one or two or three. For instance, a sporting goods manufacturer may sell two hundred products, such as baseballs, baseball bats, and tennis rackets. Each product is made up of materials like pigskin, rubber, and wood. The products made out of the same materials still differ from each other because of differing proportions. If we say that all the sporting goods are made of ten materials in varying proportions, the question is: How many units of which products should he produce to maximize his profit?

To find the answer, additional data are needed, such as the available quantities of materials, the precise proportions of each material going into each product, the unit profit of each product. *If* the outcome is fixed, i.e., the quantity of each product that will maximize the total profit exists, then that particular quantity of each product can be determined *iteratively*.

The single-valued solution to each variable (product) is found by first making as much of the most profitable product as possible, using up no more than the quantities of the materials available. The resulting profit and remaining quantities of materials are determined. The process is repeated with the second most profitable product; the resulting profit and remaining quantities of materials are determined, and so on with the third most profitable product, and the fourth, until all the materials are used up, or no more materials remain with which a single unit of a single product could be made.

In spite of the number of variables involved (one hundred thousand or more variables [products] are solvable), we are able to find the solution iteratively, because: (1) the existence of the solution is guaranteed, (2) the interaction between variables can be ignored, (3) the convergence of the successive solutions is guaranteed, and (4) the technical coefficients of production are constant and given, as are the unit profits and the capacity constraints.

The existence of the solution is guaranteed, as well as the convergence of successive solutions, by mathematical theorems. The required data input provides the values in the technical coefficients of production. The assumption that the interaction between variables can be ignored is serious. In other words, baseballs and baseball bats are two products whose demand is independent of each other, so that theoretically you could get a solution which says: Produce fifty thousand baseballs, but zero baseball bats!

Further, the statement that the technical coefficients of production are constant means that the employee makes the millionth unit in precisely the same time it took him to make the first. Additionally, he is as productive in the last hour of the workday as he is during the first. In other words, he neither tires nor learns.

Under these assumptions, the unique set of values of each variable (product) can be determined *iteratively*.

0.0 Structure-oriented Reasoning

The type of problems calling for machines with humanlike capabilities— the capabilities of seeing, reasoning, planning, and supervising—falls under the second category of our definition of reasoning: application of abstraction for *action*.

Unlike the first category, *identification*, where the outcome is fixed, in the second category, the outcome is not fixed. Furthermore, the outcome can be *changed by design*.

When the sporting goods manufacturer determined the unique value of each product that would maximize his profit, he was controlling the production technological performance capabilities of the supply variables. If he only controlled the supply variables, however, he could produce the best set of products for which there was no demand. We assumed that not only the technological performance capabilities, but also the fulfillment characteristics, were under the control of the sporting goods manufacturer. He had but to produce, and the customers would flock to his store and buy.

In the market battlefield, even as in the military battlefield, each participant controls some of the variables, but not all of the variables. The sporting goods manufacturer would control most or all the supply variables; the consumers would control most or all of the demand variables. The sporting goods manufacturer would try to influence the demand variables, through advertising, for instance, but the outcome depends not on the amount of advertising, but on the response to the advertising by the consumer.

In other words, the outcome is *interactive*. Neither side achieves the objectives fully; but both sides achieve the objectives partially.

Recalling that interaction between variables was ignored in solution-oriented reasoning, we can say that interactions are the crux of what we may call *structure-oriented* reasoning problems. In solving for the production values of each product, in the solution-oriented problem, we added the product of quantity $X_1, X_2, \ldots,$ X_n and the respective profit $C_1, C_2, \ldots, C_n$. If X_1 was the quantity of base-

balls, and X_2 the quantity of baseball bats, we know that the demand for the two interact, one representation of which interaction is X_1X_2. In the solution-oriented reasoning, we *assume* X_1X_2, or in general $X_jX_{\bar{j}}$, the product of $X_{\bar{j}}$ with all other products X_jX_j, to be 0.

0.0 Interactions in the Apollo 13 Mission

What does it mean to say that interactions are the crux of structure-oriented reasoning?

Consider Apollo 13. It will be recalled that it was one mission that had to be aborted. Things had been going quite well. In fact, the Apollo crew were quite jovial in their television appearance beamed down to earth, doing flip-flops and making wisecracks in their conversation with Houston. Right after that, the commander somberly told ground control: "Houston, we have a problem."

Complete loss of power in the spacecraft forced them to crawl over into the lunar excursion module (LEM). Houston could not identify the problem. The LEM was designed to support the few hours of activity planned for the lunar surface; it was not designed to support life on the journey. Yet LEM proved to be capable of the extra duty. The spacecraft, already beyond the halfway mark, flew to the moon and returned to earth.

On the ground, after six months of intense study, the problem was traced to a $1.98 item, a thermostat. When the thermostat malfunctioned, the rising temperature went undetected. The high temperature caused hydrogen, compressed at high pressure into liquid, to expand and knock out the power supply sources in the spacecraft.

The thermostat was one of the elements of the spacecraft. Astronaut Frank Borman, who was on both the Mercury and Apollo flights, told the author that there were twenty thousand parts in the spacecraft. By counting collection of parts functioning as a subsystem as a single part, the twenty thousand parts would be twenty thousand variables, $X_1, X_2, \ldots, X_{20,000}$.

The variables do interact; they influence their collective functioning, as so dramatically shown by the $1.98 thermostat interacting with the liquid fuel, and the liquid fuel knocking out the cabin power source.

It takes a minimum of two variables to interact: X_1X_2, X_1X_3, $\ldots$, $X_1X_{20,000}$, a total of 19,999 interactions of X_1 with the other variables.

The first variable X_j can be chosen in 20,000 ways. Corresponding to the first choice, the second variable can be chosen in 19,999 ways. When the first and second variables are already chosen, the third can be chosen in 19,998 ways, and so on. The combinations of 2 variables taken at a time from 20,000 variables is 199,999,000: let us round it to 200 million.

We can simplify the Apollo Mission as comprising two elements: the spacecraft and the astronaut. Let us further simplify by saying that the astronaut has ten thousand parts. The first order interactions of the astronaut come to approximately 50 million.

The spacecraft-astronaut unit has a total of 30,000 variables. The first-order interaction of the combined unit is *not* the sum of the interactions of the two considered separately; it is 450 million.

But interactions do not occur just in pairs. Variable X_1 could interact with both X_2 and X_3. It could interact with X_2 and X_4, and so on.

The number of interactions rises rapidly. While the first-order interactions for the spacecraft alone were 200 million, the corresponding second-order interactions rise to 1,333 billion. For the astronaut, the second-order interactions rise to 333 billion. And for the spacecraft-astronaut combination, the number of second-order interactions is approximately *4,500 billion!*

Clearly, if we were to obtain the values of all the 30,000 variables and of all the 4,500 billion interactions, the optimal trajectory selected on that basis would be invalidated by the next set of values of the variables and interactions. And even with the fastest computers, the updating of 4,500 billion values would take a significant amount of time and effort.

For one thing, we do not have the values of the 4,500 billion interactions. For another, even if we got them, the result would be outdated by the time the calculations were made. If we try to chase the interactions, the third-order interactions rise to 33,750 *trillion!*

0.0 Solution in Structure-oriented Reasoning

Analytical solution to the trillions of interactions would *not* solve the lunar-landing problem. If we imagine ourselves to be looking at the lunar-landing problem in the sixties, before Apollo 11, our solution would have to indicate the optimal approach to a totally new problem, unlike anything ever attempted, viz., returning safely to the earth the crew going on the 250,000-mile journey to the moon.

The lunar landing would not have been possible without *dynamic programming*. The notion of the solution is totally different from the one in the solution-oriented reasoning.

The solution in *solution-oriented reasoning* is a unique symbol(s) fulfilling the specified conditions and characteristics.

The solution in *structure-oriented reasoning* is the optimal sequence(s) of current resource commitment, r_i, for subsequent results, r_{j+1}, based on current results, r_j, from immediately-prior resource commitment, r_{i-1}.

The lunar-landing problem is a multi-stage decision process. From the starting point, Cape Kennedy (Canaveral), the destination moon can be reached in a number of different ways: a single-shot landing straight from the Cape to the lunar surface, a single-shot landing in a space platform, and subsequent jump onto the lunar surface, etc.

How should the multi-stage process be split into manageable segments? The critical consideration is ignorance of the nature of the journey. Since we did not know how man would perform in the suborbit, in the earth orbit, etc., let alone in

the much farther lunar orbit, *learning* about man-in-the-space-environment becomes the crucial consideration. How should the lunar journey be broken up so that the knowledge gained in one stage could be applied to the *subsequent* stage most effectively, maximizing the learning over the selected stages?

At each stage so chosen, the decision can be made to proceed or not to proceed to the next stage. We spent three missions in the earth orbit preparing for the transfer and lunar orbit missions.

How would we achieve the best results if we broke up the problem into one sequence of stages as opposed to another? The theory of dynamic programming gives us the assurance that if the Principle of Optimality is observed, the resulting functional equations system would ensure the maximum returns over *n* stages of the problem treated as *n* single-stage problems.

0.0 Structure-oriented Reasoning: (1) Interacting Variables

The lunar-landing problem is one in which the interaction of the variables is crucial. Since the analytic solution to the interactions is neither practical nor meaningful to the problem solution, what we do in dynamic programming is to *ignore interactions intelligently* by selectively studying the *empirical* results of the interaction of the variables. Which interactions were found to be important at the suborbit stage; the earth orbit stage, the transfer orbit stage, the lunar orbit stage, the lunar surface stage? By applying the learning from each stage to the subsequent stage, we avoid the necessity of considering an astronomical number of interaction values. If new problems crop up, e.g., some interactions ignored in the earth orbit stage emerge in the transfer orbit stage, corrections can be effected immediately without jeopardizing the overall mission.

As our astronomical number of interactions indicates, the interaction of the variables is what we are concerned about in the lunar-landing-type problem. Despite the enormity of its size, the problem is still one in which there is an assured endpoint, as in the helicopter search for the farmhouse. The landing location on the moon is set (within four miles of the target, as it turned out on Apollo 11). It can be reached via different sequences of decision points, but it can be definitely arrived at.

0.0 Structure-oriented Reasoning: (2) Interacting Participants

We constrasted the multitude of interactions in the lunar-landing problem with the ignoring of interactions in the sporting goods manufacturing problem. There were two types of interactions which were ignored: (1) the interaction among variables (e.g., the demand for baseball bats being related to the demand for baseballs), and (2) the interaction among participants (e.g., the seller controlling the production [supply] variables are the buyer controlling the sales [demand] variables).

The decision that both participants, the producer and the consumer, have to make focuses on a *single attribute*, viz., maximum profit or minimum cost. Game Theory assures that A, the producer, can assure himself a specific minimum value

of profit, no matter what B, the consumer does. At the same time, the consumer can assure himself a specific maximum of price, no matter what the producer does. For the MinMax Theorem to work, the payments that B makes to A have to be determined in units of profit (price) corresponding to different acts on the part of A and on the part of B.

0.0 Structure-oriented Reasoning: (3) Interacting Participants—Multi-attribute Decision

When conflicting objectives have to be reconciled, as in the case of choosing a nuclear power plant site, the participants have to focus on a *multi-attribute* utility function. The decision is based on two elements: outcome and attitude. Radiation hazard to humans is an outcome; so is the loss of salmonides due to thermal pollution. The question is not what is the *minimum* value of radiation and of salmonide loss that the community can live with. The question is what is the *maximum* value of radiation and of salmon loss that the community can live with. The interaction of the variables, such as that between biological and environmental impact, makes the *attitude* toward different levels of the variables as important as the outcomes themselves.

An additive utility function can be constructed, whose scaling constants will indicate the relative importance of changing the values from the worst to the best level of each variable. The interaction of outcome and attitude of several participants are reflected in the final choice of a nuclear power plant site.

0.0 Structure-oriented Reasoning: (4) Interacting Participants—Multiattribute, Multioutcome, Interactive Sequential Decisions

Structure-oriented reasoning concerns itself with outcomes that can be affected.

In the lunar-landing problem, the destination, the moon, remains fixed, but the *cost* in reaching the moon before 1969 would be significantly affected by the choice of the sequence of orbits. The spacecraft and astronaut *variables* interact, and the intelligent ignoring of some of the interactions is a major objective of the sequence of stages into which the problem is broken up.

In the sporting goods manufacture and sales problem, the production *quantity* would be changed by design. If the manufacturer found that the market was not responding to the demand inducements the way he would like it to respond, he would cut back on production. If, on the other hand, the market was responsive, the manufacturer would step up production. The *participants interact* and the outcome is changed.

In the nuclear plant site problem, the plant *site* would be changed by design. The interaction of variables and attitudes toward outcome is reflected in the additive utility function. The participant interaction can change the site or even abandon the construction altogether.

Instead of cost, quantity, and site changed as a result of participant interac-

tions, in the fourth situation of structure-oriented reasoning, strategic nuclear war deterrence becomes the objective outcome.

Unlike lunar-landing cost, sports goods quantity, and nuclear plant site—all of which are concrete outcomes—, the outcome in this instance is abstract: deterrence.

One way to determine the strategy in international relations is to assess the probability of moves and countermoves. Many moves in international relations are intentionally "leaked" to test the possible responses. If the reaction is unacceptable, the move can be disavowed. Since the medium often is the message in international relations, in the sense that the form of the move is as important as the content, it becomes necessary to give explicit recognition to the deliberate non- and misstatements in interactive international relations.

A *bluff* is any characterization of a situation. It becomes a *threat* when an effective response to the bluff is sought. The statement "There is a bomb in this building" becomes a threat when the people inside the building look to the nearest exit. If they do not, the statement remains a "bluff."

What effective responses are available to countries?

We can exclusively and exhaustively classify every action of a country as one or more of four categories: (1) commercial, (2) diplomatic, (3) political, and (4) military.

All Bluffs and all Threats are classifiable according to four categories. Our concern here is not with the day-to-day exchanges between countries in the form of statements and counterstatements, but with the *macrocontent* of the exchanges. When a country makes a primarily commercial move, e.g., if the U.S. banned imports of Polish ham into the United States, what would be the response? Would the commercial bluff be responded to with a commercial threat? Or would the bluff of Polish imports embargo lead to a diplomatic move by Poland, e.g., expulsion of U.S. diplomatic representatives from Poland? Would the diplomatic threat by Poland lead to a response in kind, or something graver, e.g., a U.S. effort to destabilize the Polish government, which would be a political move? Again, would the Polish response be in kind, e.g., influencing the outcome of presidential elections, or something graver, e.g., declaring war—conventional war—on the United States? In that case, the political move would have been responded to with a military move.

The purpose of this highly hypothetical scenario is to suggest that bluffs become threats only when a positive response is made in kind, or in one of the other three categories of moves. Every move that every country makes at any time must be classifiable into one or more of the four categories. A comprehensive scheme of 110 "Bluffs" and 90 "Threats" has been developed in the Bluffs-Threats Matrix, into one or more of whose cells every international relations move is classifiable.

The interactive behavior of the participants is characterized by the transition probabilities from one state (move) to another. The possible transitions are: (1)

Figure 7.1
Illustrative Interactive Sequence

STATE S(0)
U.S.S.R. made a POLITICAL MOVE
U.S.A. responded with a DIPLOMATIC MOVE

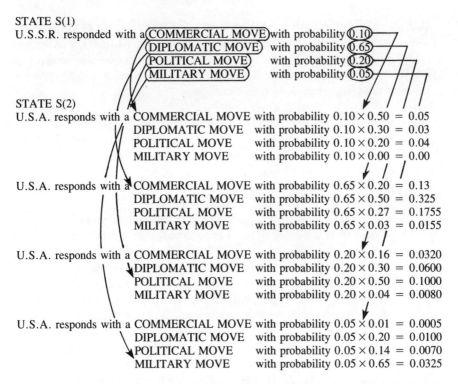

STATE S(1)
U.S.S.R. responded with a COMMERCIAL MOVE with probability 0.10
                          DIPLOMATIC MOVE with probability 0.65
                          POLITICAL MOVE   with probability 0.20
                          MILITARY MOVE   with probability 0.05

STATE S(2)
U.S.A. responds with a COMMERCIAL MOVE with probability $0.10 \times 0.50 = 0.05$
                       DIPLOMATIC MOVE with probability $0.10 \times 0.30 = 0.03$
                       POLITICAL MOVE   with probability $0.10 \times 0.20 = 0.04$
                       MILITARY MOVE   with probability $0.10 \times 0.00 = 0.00$

U.S.A. responds with a COMMERCIAL MOVE with probability $0.65 \times 0.20 = 0.13$
                       DIPLOMATIC MOVE with probability $0.65 \times 0.50 = 0.325$
                       POLITICAL MOVE   with probability $0.65 \times 0.27 = 0.1755$
                       MILITARY MOVE   with probability $0.65 \times 0.03 = 0.0155$

U.S.A. responds with a COMMERCIAL MOVE with probability $0.20 \times 0.16 = 0.0320$
                       DIPLOMATIC MOVE with probability $0.20 \times 0.30 = 0.0600$
                       POLITICAL MOVE   with probability $0.20 \times 0.50 = 0.1000$
                       MILITARY MOVE   with probability $0.20 \times 0.04 = 0.0080$

U.S.A. responds with a COMMERCIAL MOVE with probability $0.05 \times 0.01 = 0.0005$
                       DIPLOMATIC MOVE with probability $0.05 \times 0.20 = 0.0100$
                       POLITICAL MOVE   with probability $0.05 \times 0.14 = 0.0070$
                       MILITARY MOVE   with probability $0.05 \times 0.65 = 0.0325$

Commercial-Commercial (CC), (2) Commercial- Diplomatic (CD), (3) Commercial-Political (CP), and (4) Commercial-Military (CM), when the initial move is Commercial. Similarly, PC, PD, PP, and PM are another set of Bluff-Threat transitions.

Since the outcome devoutly desired to be avoided is (Row 110, Column 90), viz., strategic nuclear war "bluff" responded to by strategic nuclear war "threat," we can develop a metric of the "distance" from any given move to the initiation of strategic nuclear war. While deterrence is at best stated probabilistically, it can be measured numerically. For instance, if a "commercial move" is made against a country like Poland, what is the number of steps in an escalation to strategic nuclear war? Such a metric has been developed and applied to illustrate "Bluff-Threat" scenarios. Fig 7-1 illustrates an illustrative *interactive* sequence.

What AI research is expected to yield by 1989 is providing computers with

181

humanlike capabilities: seeing, reasoning, planning, and supervising. We have shown that the goal will be met by providing reasoning as we define it: The process of abstraction of, and/or application of abstraction to characteristics of entities identification and/or action. *Iterative* methods would generally apply to identification, while *interactive* methods would generally apply to action.

0.0 Identification and Action Capabilities of Expert Systems

Given that DARPA with its billion-dollar commitment to AI research in the next seven years, expects the demonstration of reasoning capabilities that would both identify entities and act to change outcomes, where do the present-day expert systems stand with respect to those twin capabilities?

0.0 Defining the Expert System

Building Expert Systems is the first volume of the *TeKnowledge Series in Knowledge Engineering*. It is the fruit of a unique collaboration of thirty-eight top expert systems researchers and developers.

Donald Waterman and Frederick Hayes-Roth offer the following definition:

> An expert system is a computer program that embodies the expertise of one or more experts in some domain and applies this knowledge to make useful inferences for the user of the system. More specifically, an expert system has a number of basic characteristics that distinguish it from conventional programs. First, it must perform well. . . . Second, it bases its reasoning process on symbol manipulation. . . . Third, an expert system must have some knowledge of the basic principles of its domain of interest. . . . Fourth, it must be able to reformulate the problem . . . into an internal representation convenient for processing with its expert rules. Finally, the expert system must handle formidable problems in a real-world domain.[4]

The "user of the system" is the one who should find the inferences made by the expert system "useful." One of the most celebrated uses of the expert system is provided by PROSPECTOR, which, according to John Gaschnig of SRI International, has discovered a molybdenum deposit whose ultimate value will probably exceed $100 million.[5] The user of this information is the corporation that will

4. Donald A. Waterman and Frederick Hayes-Roth, "An Investigation of Tools for Building Expert Systems," in *Building Expert Systems*, ed. Frederick Hayes-Roth, Donald A. Waterman and Douglas B. Lenat (Reading, Mass.: Addison-Wesley, 1983), 169–70.
5. *Ibid.* 6.

dig for the deposits. It is a management decision to commit with uncertainty resources today for results tomorrow.

Uncertainty is not ignorance of general outcome, but of specific outcome. In fact, it is certain that one event/object/relationship out of a group of events/objects/relationships will occur; but *which* one will occur is *unknown* and *unknowable* in advance.

Whether or not the molybdenum deposit will yield $100 million is not known today; and it is not knowable in advance. The results of the digging for molybdenum may range from a yield of nothing to $100 million or more.

Given unknown and unknowable specific outcomes in the future, an *action path* has to be selected. The action path is the sequence of *predetermined* choices between potential alternatives. In every choice, and sequence of choices, there is *risk*, the possibility of the occurrence of an outcome other than the one specified. Given the same risk, the action will differ, depending on the decision-maker's *attitude*, the readiness to accept adverse (or favorable) outcomes.

PROSPECTOR can assist the decision-maker in choosing an action path with reduced uncertainty. The reduction in uncertainty comes from the expertise of PROSPECTOR.

0.0 What Is Expertise?

Brachman points out that the experts work with very sketchy rules. Their rules are high-level rules that reduce the search area quite efficiently. While they quickly eliminate many possibilities in each inferential move by using high-level rules, they are able to deal with the unexpected patterns by falling back upon reasoning from first principles. General problem-solving abilities permit the expert to make conceptual extrapolations from, as well as interpolations between, data and high-level rules:

> The ability to fall back on reasoning from first principles
> is clearly one quality that sets expert systems apart from
> Fourier transform programs. Work in this field to date,
> however, has emphasized finding the macromoves of
> expert-level inference, so that most fundamental knowl-
> edge has been ignored.[6] Yet the intent of expert systems
> work is to produce systems that are as robust in their
> fields as human experts are. Thus, while not yet a com-
> mon property, general problem-solving ability in the do-
> main of expertise, coupled with extensive knowledge of
> basic principles, is an important axis along which to mea-
> sure expert systems.[7]

6. P. E. Hart, "What's preventing the widespread use of expert systems?" Position paper, Expert Systems Workshop, San Diego, CA., 1980, 11–14.

7. Ronald J. Brachman, Saul Amarel, Carl Engelman, Robert S. Engelmore, Edward A. Feigenbaum, and David E. Wilkins, "What are Expert Systems?", in *Building Expert Systems, op. cit.*, 47.

Expertise applies to both solution-oriented and structure-oriented reasoning. The expert finds the unique symbol(s) fulfilling the specified conditions and characteristics. When the path to the solution is direct, the situation is *not* worthy of an expert. When the path to the answer is not obvious, however, and/or when the situation admits multiple solutions, the expert ensures the best solution.

Multiple solutions suggest structure-oriented reasoning. Interactions are numerous, and they are important. Most interactions, however, have to be ignored in a complex problem. Expertise is critical to the selection of the interactions to be ignored.

How is the expert able to ignore interactions intelligently? From experience of prior instances of complexity; from observing successful practitioners; from knowing what the domain of expertise holds as collective wisdom, distilled from trial, error, and experience.

In other words, expertise is the efficient application of the general to the particular. It is precise, prompt, and adaptive. The quick answer of the expert will be the same as one arrived at by a much larger number of steps. The expertise shines when unexpected difficulties arise; alternative approaches are devised almost instantaneously. The expert processing is adaptive. His adaptive methods are made possible by his depth of knowledge of his own field, as well as by the applicability of problem-solving methods.

We can define expertise as *an accurate and adaptive application in a specified domain of knowledge, of the general to the particular, to identify symbol(s) fulfilling specified conditions and characteristics, and/or optimal sequence(s) of current resource commitment for subsequent results, most accurately, most often, with least delay.*

Clearly, "most accurate" is a relative term. DDT was hailed for the "Green Revolution" until 1952, when the gas spectrochemographic technique was introduced and made it possible to detect many substances in concentrations of one part per million or less. The technique revealed the previously unsuspected fact that DDT was unlike other most complex chemical compounds in that it was highly resistant to degradation.[8] The agricultural expert advising massive doses of DDT to increase crops would be "most accurate" until 1952 when the domain of his expertise was significantly modified. As long as he had been "most accurate, most often" in rendering expert advice, he would have faithfully discharged his professional obligation.

How do we incorporate the concept of expertise into the definition of expert systems? To say that expert systems are those that embody expertise of experts does not say much about what constitutes expertise. The cynic would say that the expert is someone from out of town. When do we know that the expert systems embody the expertise of one or more experts? Given the same data, we must de-

---

8. National Goals Research Staff, *Toward Balanced Growth: Quantity with Quality*, Government Printing Office, Washington, D.C., 1970, 63.

termine the advice of experts in person and compare the advice of the expert system with that of the experts. This is, of course, more easily said than done. How the expert thinks is not often known to him, because the intuitive leaps he makes accurately are second nature to him. If the expert system does not capture all of the hidden reasoning processes, its answer would fall short of the expert's answer. And it would not be fair to compare the two results. It makes it all the more necessary to label the resemblance of the computer to the human expert.

0.0 *"Usefulness to the User"*

What inference becomes "useful" to "the user of the system"? If the inference is certain, e.g., $2 + 2 = 4$, the user does not need an expert. But if the "2" on the left and the "2" on the right are not identical quantities, and if the addition would result in mixing apples with oranges, then we would need the in-depth knowledge of someone who knows what the symbols really represent, as well as the reasonable combining of the two symbols.

Why would the user want to know what the combined result is? Invariably, to reduce the uncertainty. The outcome of combining "2" and "2" must enable the user to commit more resources, less resources, or no resources or defer the commitment decision until additional data are collected and interpreted. Since any commitment of resources would entail risk, the possibility of the occurrence of an outcome other than the one specified, the expert system useful to the user must reduce his risk.

No expert, and therefore no expert system, can be one hundred percent accurate. PROSPECTOR will identify molybdenum deposits when there are indeed molybdenum deposits. But it will sometimes identify deposits when there are none. Further, it will sometimes identify no deposits when indeed there are positive deposits. Again, PROSPECTOR would correctly identify zero deposits when there are zero deposits.

"Useful inferences for the user" of the expert system cannot be indicated without reference to the *wrong decisions* that both the expert and the expert system will make. Witness, for instance, the parade of experts in a court of law, half of them arguing that the defendant was mentally competent, half of them arguing precisely the opposite, though no expert was present at the scene of the crime. To err is indeed human; and the judicial system recognizes that it will err. The system therefore sets up safeguards so as to err in the more preferred direction in the long run.

We can present in the same framework the four types of action based on advice from the expert system PROSPECTOR and the four types of action based on advice from the expert system CADUCEUS, which diagnoses diseases. In both instances, we need to establish the error that we want to avoid. Is it worse to dig for molybdenum when there is none, or not to dig for molybdenum when there is some? Again, is it worse to identify a disease when it is not present, or to identify zero disease when in fact there is disease?

185

The truth is not known at the time of the decision. When using PROSPEC-TOR and CADUCEUS, we know the truth only after the decision is made to dig or to operate. Upon digging we find either that there is a deposit, or that there is none. Similarly, upon operating on the patient, we find either that there is disease, or that there is none.

The track record of the expert and the expert system is established open-endedly. While the particular expert may practice his expertise for only a finite number of years, the system of expertise continues. Therefore, we apply the premise of statistics that deals with repeated phenomena. Statistical theorems are for the most part based on probability, defined as the number of specified occur-rences, divided by the total number of occurrences, as the number of trials tends to infinity. Since infinity is beyond anything finite, even a daily occurrence like the sunrise cannot be considered to have reached infinite occurrences, but only tends to infinity.

Will the sun rise tomorrow? Although mankind has witnessed a sunrise every day, the observations are not "infinite", so we can only say that the probability of sunrise tomorrow is given by probability $p = .9999$, but not 1. For decision-making purposes, we postulate the opposite: viz, the sun will not rise tomorrow, or the sunrise $= 0$. On the basis of evidence, we *reject* the (sunrise $= 0$) state-ment with 99.999999% probability.

If we consider it worse not to dig when there is molybdenum, we would want to disprove the opposite: viz., there is no molybdenum. Rejecting the true is a type I error, more serious than accepting the false, a type II error. For us, the truth is the $100 million molybdenum deposit. We postulate its opposite, viz., there is 0 molybdenum deposit, and hope to disprove it with 95%, 99%, etc. prob-ability. The statement that something is zero is called the Null Hypothesis. Our Null Hypothesis (NH) is: Molybdenum Deposit $= 0$.

The truth could be that the NH is correct, the deposit is indeed 0. Or the truth could be that the NH is wrong, i.e., the deposit is $100 million worth.

How does the user *perceive* the truth? When he perceives the 0 deposit as 0, or the $100 million deposit as a $100 million deposit, that is fine. The trouble is when his perceptions are opposite the truth. Since *we* set up the NH, we should reflect the user's attitude toward the error of not digging when there was a $100 million deposit, or digging when there was 0 deposit.

The expert system should enable the user to see the error consequences under opposing null hypotheses. The *user* decides which of the two is the worse conse-quence to him. The expert system should set forth both possibilities.

Getting back to the track record of the expert system, the user should be able to specify his preferred frequency of type I error, and know which way the expert system is likely to err: Does it *implicitly* work from a Null Hypothesis that the de-posit is 0, or that it is a positive quantity?

Table 7.1 sets forth the types of errors in decision-making with PROSPEC-TOR and CADUCEUS.

Table 7.1
Types of Errors in Decision-making When Using Expert Systems PROSPECTOR and CADUCEUS

PROSPECTOR

Perception \ Truth	Y		N	
	Truth	Perception	Truth	Perception
Y	$0	$0	$0	$100 mil.
N	$100 mil.	$0	$100 mil.	$100 mil.

Type I Error

NH: Molybdneum deposit = 0

Perception \ Truth	Y		N	
	Truth	Perception	Truth	Perception
Y	$100 mil.	$100 mil.	$100 mil.	$0 mil.
N	$ 0 mil.	$100 mil.	$0 mil.	$0 mil.

Type I Error

NH: Molybdenum deposit = 100 OR, (Moly.dep. −100) = 0

CADUCEUS

Perception \ Truth	Y		N	
	Truth	Perception	Truth	Perception
Y	no disease	no disease	no disease	disease
N	no disease	disease	disease	disease

NH: Disease = 0

Perception \ Truth	Y		N	
	Truth	Perception	Truth	Perception
Y	disease	disease	disease	no disease
N	no disease	disease	disease	disease

Type I Error

NH: There is disease, d OR (disease − d) = 0

0.0 Redefining Expert Systems

How can we incorporate our definition of expertise and the elements of usefulness to the user of the expert system in redefinition? *The expert system is a computer-based aid, including algorithms and interfaces, which replicates the accurate and adaptive application in a specified domain of knowledge, of the general to the particular, by one or more experts, to identify symbol(s) fulfilling specified conditions and characteristics, and/or optimal sequence(s) of current resource commitment for subsequent results, most accurately most often with least delay.*

As the field evolves, so will the definition of expert systems. In this redefinition, we emphasize the replication of the human expert as the test of embodiment of expertise. The embodiment, if properly carried out, will replicate the original, viz., the expert(s). We want to define the expert system apart from the use or misuse by the user, but we want to highlight the unique contribution of its knowledge and experience base. By stating that the computer-based expert system replicates the human originals most accurately, most often, and with least delay, we emphasize the contribution of the expert system, leaving it to the user to make the best use of it.

While the expert system generates results in both the solution-oriented reasoning situations and structure-oriented situations, so far the concentration has been in the former. The authors of *Building Expert Systems* hope their book to be "useful as a text for computer-science and decision-support system students." How are decision support systems different from expert systems? One difference is orientation. Instead of the solution-oriented emphasis of the expert system, the DSS aims to be structure-oriented. We define DSS in chapter 12: *The decision support system (DSS) is basically a computer-based aid, including algorithms and interfaces, primarily to assist the user to commit resources to achieve results in situations with little prior applicable experience, by his(her) progressively discovering alternative resource-results combinations corresponding to changing priorities of objectives.*

0.0 Expert Systems Accenting Identification

We can now look at the major expert systems that are discussed in the book *Building Expert Systems*.

Most of the expert systems focus on what we have called identification, which is determining the unique symbol(s) fulfilling the specified conditions and characteristics.

DENDRAL infers molecular structure from mass spectrographic information.[9] The oldest expert system, DENDRAL began in 1965 and uses a specialized data structure that allows the efficient retrieval of frequently used facts and the

9. B. G. Buchanan, G. L. Sutherland, and E. A. Feigenbaum, "Heuristic DENDRAL: A Program for Generating Explanatory Hypotheses in Organic Chemistry," in *Machine Intelligence*, ed. B. Meltzer and D. Michie, Vol. 4, Edinburgh: Edinburgh University Press, 1969, 209–54.

storage together of facts used together.[10] Depth-first, instead of breadth-first, search yields quick hypotheses on the molecular structure. There is a structure-proposing program in DENDRAL which generates complete and nonredundant candidate molecular structures faster and more accurately than human chemists.[11]

Turning from the chemical expert system DENDRAL to the medical expert system CASNET (Causal Associational Network Program), the output can still be characterized, in our terminology, as identification. DENDRAL uses static rules, as does MYCIN, which gives consultative advice on diagnosis and therapy for infectious diseases. CASNET, however, which was developed during the early and mid-seventies, accomodates probabilistic rules,[12] as does CADUCEUS:

> Reasoning under uncertainty is an important part of the approach used by CADUCEUS. The solution method combines data-directed and hypothesis-directed data interpretation under constraints of disease taxonomy and causality relationships. Knowledge in the system is represented as a network of findings and diseases via varied LISP programs. Knowledge acquisition involves editing the links between disease entities and the findings or symtpoms.[13]

CADUCEUS builds disease models from its knowledge base of 500 diseases, 350 disease manifestations, and 100,000 symptomatic associations, and partitions the disease tree into diseases corresponding to patient symptoms.

Developed around the same time as CASNET, MYCIN also uses probabilistic reasoning on infectious diseases.[14,15] It assesses propositions on a scale from -1 (certainly wrong) to $+1$ (certainly right). MYCIN allows propositions to be partially right or wrong, unlike in the predicate calculus. Consider the following example of a rule from MYCIN:

If   the infection is primary-bacteremia, and the site of the culture is one of the sterile sites, and the suspected portal of entry of the organism is the gastrointestinal tract,

then   there is suggestive evidence (0.7) that the identity of the organism is bacteroids.

10. *Building Expert Systems, op. cit.*, 120.

11. *Ibid.*, 44

12. S. M. Weiss, C. A. Kulikowski, and A. Safir, "A Model-based Consultation System for the Long-term Management of Glaucoma," *Proceedings of the 5th International Joint Conference Artificial Intelligence*, 1977, 826–32.

13. Ronald J. Brachman, Saul Amarel, Carl Engelman, Robert S. Engelmore, Edward A. Feigenbaum and David E. Wilkins, "What are Expert Systems?" in *Building Expert Systems*, op. cit. 54.

14. E. H. Shortliffe, *Computer-based Medical Consultation: MYCIN* (Elsevier, New York: 1976).

15. R. Davis, B. G. Buchanan, and E. Shortliffe, "Production Rules as a Representation for a Knowledge-based Consultation Program," *Artificial Intelligence*, 8:15–45.

The number 0.7 in this rule indicates that the evidence is strongly indicative, but not absolutely certain. Evidence confirming a hypothesis is collected separately from that which disconfirms it, and the "truth" of the hypothesis is the algebraic sum of the evidence.[16]

Turning from the medical expert systems to *geological expert systems*, PROSPECTOR, developed in 1979, is structurally similar to MYCIN.[17] Like MYCIN, PROSPECTOR determines the degree of support for each antecedent condition from the data. The Knowledge Acquisition System (KAS) facilitates the acquisition of all kinds of knowledge, prompting the user continually until all missing parts of a new structure are filled in. PROSPECTOR's prediction of a molybdenum deposit in a location was confirmed by a $100 million find.

0.0 From the Iterative to the Interactive

The iterative process of zeroing in on the unique symbol(s) fulfilling the specified conditions and characteristics works well with solution-oriented problems. But the severe limitation imposed on the problems includes the ignoring of interactions among participants, and even interactions among variables. The celebrated PROSPECTOR, structurally similar to the prolific MYCIN (inspiring EMYCIN, PUFF, PROSPECTOR, and, to some extent, TEIRESIAS), cumulates *algebraically* the evidence both in support of the hypothesis and against it. It is similar to the maximization of total profits in the sporting goods manufacture problem, where the contribution to profit by each product is considered independently of all the others. The real-life interaction among the variables, e.g., baseball and baseball bat, is ignored.

We have identified four types of important real-life problems. In solving each of these problems, the *interactive* approach, as opposed to the iterative approach, is critical, whether:

- the decision-making centers around the allocation of resources for results in the subsequent stage, based on the (results-resources) relationship of the preceding stage (dynamic programming);
- or the decision-making centers around the mixture of maximum problems, each participant trying to maximize his outcome but having to settle for less, since the participants control only some of the determining variables some of the time (game theory);
- or the decision-making centers around balancing conflicting objectives of multiple interests in a collective project (multiattribute utility function);
- or the decision-making centers around open-ended policy moves in different dimensions, such as commercial, diplomatic, political, military, the *perceptions* of policy being as important as the policies themselves, if not more so,

16. *Ibid.*
17. R. O. Duda, P. E. Hart, P. Barrett, J. Gaschnig, K. Konolige, R. Reboh, and J. Slocum, "Development of the PROSPECTOR Consultation System for Mineral Exploration," Final Report of SRI Projects 5821 and 6415, SRI International, Menlo Park, CA. 1978.

with an abstract outcome, such as deterrence, being the objective (bluffs-threats matrix and aggregate outcome transitions).

0.0 Interactive Blackboard for Distributed Problem-solving

There appears to be a hardware-induced possibility for realizing interactive processes for problem-solving, eventually leading to the action-type outcomes of the structure-oriented problems discussed in the last section.

Victor Lesser and Daniel Corkill suggest that the improved capabilities in and experience with Distributed Processing Systems enable us to think seriously of *Distributed Problem-solving* (also called *Distributed AI*):

> We broadly define distributed problem solving networks as distributed networks of semi-autonomous problem solving *nodes* (processing elements) that are capable of sophisticated problem solving and cooperatively interact with other nodes to solve a *single* problem. Each node can itself be a sophisticated *problem solving system* that can modify its behavior as circumstances change and plan its own communication and cooperation strategies with other nodes. . . .
>
> Networks of cooperating nodes are not new to artificial intelligence. However, the relative autonomy and sophistication of the problem solving nodes, a direct consequence of limited communication, sets distributed problem solving networks apart from most others, including Hewitt's work on the actor formalism,[18] Kornfield's ETHER language,[19] Lenat's BEING system,[20] and the augmented Petri nets of Zisman[21]. . . .
>
> Most initial work in distributed problem solving has focused on three distributed air traffic control, and distributed robot systems.[22,23,24] All of these applications

18. C. Hewitt, "Viewing Control Structures as Patterns of Passing Messages," *Artificial Intelligence*, Vol. 8, No. 3, 1977, 323–64.

19. W. A. Kornfield, "ETHER: A Parallel Problem-solving System," *Proc. 7th Int. Jt. Conf. Artificial Intelligence*, 1979, 490–92.

20. D. B. Lenat, "BEINGS: Knowledge as Interacting Experts," *Proc. 3rd Int. Jt. Conf. Artificial Intelligence*, 1975, 126–33.

21. M. D. Zisman, "Use of Production Systems for Modeling Asynchronous, Concurrent Processes," in *Pattern-Directed Inference Systems*, ed. D. A. Waterman and Frederick Hayes-Roth (New York: Academic Press, 1978).

22. R. Davis, "Report on the Workshop on Distributed Artificial Intelligence," *SIGART Newsletter*, 1980, 43–52.

23. R. Davis, "Report on the Second Workshop on Distributed Artificial Intelligence," *SIGART Newsletter*, 1982, 13–23.

24. M. Fehling and L. Erman, "Report on the third annual workshop on Distributed Artificial Intelligence," 1984, 3–12.

need to accomplish distributed interpretation (situation assessment) and distributed planning. Planning here refers not only to planning what actions to take (such as changing the course of an airplane), but also to planning how to use resources of the network to carry out the interpretation and planning task effectively. This latter form of planning encompasses the classic focus-of-attention problem in AI.[25]

Lesser and Corkill chose Distributed Vehicle Monitoring as the ideal problem to pursue Distributed Problem-solving. A number of processing nodes and associated acoustic sensors are distributed geographically over a wide area. The acoustic sensor has limited range and accuracy and can identify nonexistent vehicles as well as miss real vehicles. This possibility of error, and the need to communicate with proximate nodes to identify the vehicle, make it appropriate to view distributed vehicle monitoring as a Functionally Accurate, Cooperative (FA/C) System. In FA/C, "the distributed system is structured so that each node can perform useful processing using incomplete input data while simultaneously exchanging the intermediate results of its processing with other nodes to construct cooperatively a complete solution."[26]

Central to the solution of the Distributed Vehicle Monitoring is the concept of the Blackboard model developed for HEARSAY-II speech-understanding system.[27] Spoken utterances suggested hypotheses on the basis of partial knowledge: acoustic, phonetic, syllabic, lexical, syntactic, and semantic knowledge. HEARSAY-II used a global working memory, the *Blackboard*, on which the different knowledge sources (KS) could "write" their hypothesis of what the utterance is. HEARSAY-II would shift attention appropriately from one area of the interpretation problem to another. When a KS creates a hypothesis from previously created hypotheses, the KS extends the existing (partial) interpretation, thereby reducing the uncertainty of the overall interpretation. Processing terminates when a consistent hypothesis is generated that satisfies the requirements of a complete solution.

In FA/C, the basic HEARSAY-II architecture has been extended. Goal-directed control has been integrated into HEARSAY-II's data-directed control. A goal blackboard and a meta-level control blackboard have been added, in addition to communication knowledge sources and a planning module:

25. Victor R. Lesser and Daniel D. Corkill, "The Distributed Vehicle Monitoring Testbed: A Tool for Investigating Distributed Problem Solving Networks," *The AI Magazine*, Vol. 4, No. 3, Fall 1983, 15, 16.

26. Victor R. Lesser and Daniel R. Corkill, "Functionally Accurate, Cooperative Distributed Systems," *IEEE Trans. Sys., Man, and Cybernetics*, SMC-11, 1, Jan. 1981, 1982.

27. L. D. Erman, F. Hayes-Roth, V. R. Lesser and D. R. Reddy, "The HEARSAY-II Speech Understanding System: Integrating Knowledge to Resolve Uncertainty," *Computing Surveys*, Vol. 12, No. 2, 1980, 213–53.

The *goal blackboard* mirrors the structure of the data blackboard. Instead of hypotheses, the basic data units are *goals*, each representing an intention to create or extend a hypothesis with particular attributes on the data blackboard. For example, a simple goal would be a request for the creation of a vehicle location hypothesis above a given belief in a specified area of the data blackboard.

Goals are created on the goal blackboard by the *blackboard monitor* in response to changes on the data blackboard. These goals explicitly represent the node's intention to abstract or extend particular hypotheses. Goals received from another node may also be placed on the goal blackboard. Placing a high-level goal onto the goal blackboard of a node can effectively bias the node toward developing a solution in a particular way.

The *planner* responds to the insertion of goals on the goal blackboard by developing plans for their achievement and instantiating knowledge sources to carry out those plans. The *scheduler* uses the relationships between the knowledge source instantiations and the goals on the goal blackboard as a basis for deciding how the limited processing and communication resources of the node should be allocated.[28]

We concluded chapter 6 by citing Feigenbaum's view that the most powerful of the control structures for search is the blackboard model. The Distributed Vehicle Monitoring Testbed offers a facility to simulate different degrees of sophistication in problem solving knowledge using the blackboard model. The *interactive* cooperation to solve a single problem offers the hope that AI can someday tackle the important problems of interacting variables and interactive participants in decision-making.

28. Victor R. Lesser and Daniel D. Corkill, "The Distributed Vehicle Monitoring Testbed," *op. cit.*, 25.

BOOK TWO

Concomitant Coalitions in Cars and Computers

PART IV

Concomitant Coalitions in Japan and in the United States

8

CONCOLs in Trade and Technology

OVERVIEW: Importing high technology *from* the US and Japan in order to export high-tech products *to* them places Korea (and other similar trade partners) in opposing roles: Korea *with* the US (paying more for high-tech), and simultaneously *against* the US (demanding more for high-tech products exported). This association of the same party with and against other party(ies) in the same game is called Concomitant Coalition (CONCOL).

To set the stage for CONCOL in trade and technology, robotics and artificial intelligence, which are respectively the *metal* and the *mental*, are introduced as part of *adaptive automatics*. A robot is essentially an arm which can move in many directions. It can be reprogrammed—a robot digging ditches one minute can be welding metals in the next by a change of programs. Imbedding in the machine the intelligence of human problem-solving is a concern of artificial intelligence. As robots become imbued with more intelligence, they should be able, not only to merely fulfill goals, but also, to develop goals for the product (output) and design appropriate methods to precision-produce them. A First Generation ADAPTOMATON is envisaged as a self-discovering and self-designing system that transforms inputs into outputs progressively more efficiently to accomplish designed and/or discovered objectives. The recent experimental evidence of the elemental bit of genetic information, the lambda, repressing the reading of the DNA instructions, is an excellent example of adaptive automatics, operating elegantly in nature.

Turning from the possible profile of future ADAPTOMATONs to the pragmatic question of trade and technology strategy, we argue that CONCOL in trade and technology should have as its measure something other than a list of items traded. Instead, the concept of PSPG, the ratio of Performance-Specific/Performance-General, is proposed as an index of the mix of products which are high-tech in the export portfolio. Generally, the developing nations will be the customers for the performance-general outputs, and the advanced nations the customers for the performance-specific outputs. Once PSPG is accepted as a criterion, its

199

present value and potential value in the 1990s can be determined. To progress from the actual to the ideal value of PSPG is the concern of the CONCOL bargaining *process*, to prepare for which, in chapters 5 and 6, we identify the CONCOL *players* in the productivity game in Japan and the robotics game in the U.S.

0.0 Adaptive Automatics (AA), a New Concept Subsuming Robotics and Artificial Intelligence

The coming Robotics Revolution (RR), like the continuing Industrial Revolution (IR), serves to multiply inputs manyfold. Much of mankind has moved from muscle power to machine power, thanks to the Industrial Revolution, which multiplies musclepower manyfold. The coming Robotics Revolution promises to multiply musclepower even more, reliable robots working twenty-four hours a day virtually unattended by humans.

0.0 Mindpower, the Hallmark of Artificial Intelligence

A robot is essentially a hand which can move in several directions. When robot vision augments the motion of the hand, a new dimension of usefulness will be inaugurated.

But the *metal* (the robot) is only as good as the *mental* (artificial intelligence) that programs the activities of the robot. The Czech root of robot, *robota*, means enforced labor or slave labor, a slave doing what he is told without question or complaint. But who tells the robot, the modern slave, what to do? The program, or better, the programming process. The same robot digging ditches one moment could be welding metals the next moment by changing the program that runs it. This reprogrammability is emphasized in the formal definition of the robot by the Robot Institute of America: "A robot is a *reprogrammable*, multi-functional *manipulator* designed to move material, parts, tools, or specialized devices through variable programmed motions for the *performance of a variety of tasks*."[1]

The purpose of the program is to let the machine mimic the man. To accomplish this, it is essential to know how humans think. A problem that has fascinated mankind, human problem-solving holds the key to the metal and the mental. The field of study that will let us do more with the metal because we know more about the mental is artificial intelligence.

In 1956 the first conference on artificial intelligence was held at Dartmouth. John McCarthy coined the term "artificial intelligence" as the name for the conference. He pointed out that the critical missing element in making intelligent machines is neither technical nor mechanical knowledge, but knowledge about knowledge itself:

1. Robot Institute of America, Dearborn, Mich.

200

I think it's a lack of understanding of the relationship between problems and the methods for solving them. It is a
question of understanding what is known about the world,
what people know about the world, and how one can express it in terms which machines can use.[2]

At the Dartmouth Conference, the group from Carnegie-Tech (now Carnegie-
Mellon) advanced the view that basic to all human problem-solving were a few
clever methods. They succeeded in proving their proposition with reference to
theorem-proving by their Logic Theory Machine, which found proofs for thirty-
eight out of fifty-two theorems in *Principia Mathematica* that it tried.
The bold forecasts of intelligent machines within a decade by one of the cre-
ators of the Logic Theory Machine have yet to be fulfilled more than a quarter
century after the forecast:

It is not my aim to surprise or shock you. But the sim-
plest way I can summarize is to say that there are now in
the world machines that think, that learn, and that create.
Moreover, their ability to do these things is going to increase rapidly until—in a visible future—the range of
problems they can handle will be coextensive with the
range to which the human mind has been applied. [Within
ten years] a digital computer will be the world's chess
champion . . . a digital computer will discover and prove
an important new mathematical theorem . . . theories in
psychology will take the form of computer programs.[3]

0.0 Rejection of Data in Human Problem-solving
While the machines themselves have not yet appeared, we now know more about
how humans solve problems. Herbert Simon and Allen Newell devoted seventeen
years to the study of human problem-solving in chess, cryptarithmetic, and logic.
These problems are well-structured in the sense that the many possible paths are
enumerable. In chess, there are about 120 different sequences of moves. If there
were 130 different sequences, we could use the calculations made elsewhere for
the number of banana splits that could be made from 130 flavors of ice cream.[4]
The number of moves with 130 sequences is 6,467 followed by 216 zeros. If it
takes only one-billionth of a *second* to consider a move, the 31.536 million seconds in a *year* will consider a total of (3.1536 followed by 16 zeros) moves. To

2. Philip J. Hilts, *Scientific Temperaments: Three Lives in Contemporary Science* (New York: Simon and Schuster, 1982), 231.
3. Herbert Simon and Allen Newell, *Operations Research*, Vol. 6, No. 1, Jan.–Feb. 1958.
4. George K. Chacko, *Applied Operations Research/Systems Analysis in Hierarchical Decision-
making, Vol. II*, (Amsterdam, The Netherlands-North-Holland, 1975), 43.

determine the number of *years* of computation it would take to consider the possible chess moves with 130 different sequences, we have as the numerator 6.467 followed by 219 zeros, and as the denominator 3.1536 followed by 16 zeros. It would take *28.0686 years!*

Even if 28.0686 years of continuous computer consideration of chess moves at the rate of one billion a second were expended, that would by no means ensure the best, or even a passable, game of chess. The reason is that a large number of moves is irrelevant. As soon as the opponent makes a move, a whole array of moves that were theoretically possible until that moment disappears from realistic consideration. When considering all possible sequences, the machine is using computational power (brawn) without discrimination (brains).

At the Dartmouth Conference, McCarthy contributed the *alpha-beta heuristic* in chess. Instead of looking at all possible moves, look for one's most promising move and identify the most damaging response to it that the opponent can make. Eliminate from further consideration all moves that one can make which could give one's opponent a strong showing. Says McCarthy:

> Much of our intelligent behavior is like this—it is not determined by heredity, or by culture, but is shaped by the difficulties of the problem to be solved. Many linguists and psychologists have not grasped this idea yet. But it is what artificial intelligence is—studying the relationship between problems and the methods for their solution.[5]

Simon and Newell studied how experts solved the chess, cryptarithmetic, and logic problems. The players were asked to articulate why they chose certain actions and why they rejected others. From a large number of explications of the method of problem-solving, Simon and Newell developed a protocol of a few steps, e.g., 16 steps for the cryptarithmetic problem.

The Simon-Newell protocols in chess tell us in effect how McCarthy's alpha-beta heuristic is developed by different people, and how they all have certain shared characteristics. It would appear reasonable to say that human beings solve problems, not by accepting all the data offered, but, equally, if not more importantly, by *rejecting* sometimes major segments of the data offered. The alpha-beta heuristic, for instance, *rejects* all moves that can potentially provide one's opponent with seriously damaging responses.

While the rules of rejection are reasonably straightforward in chess, e.g., avoid entrapment of one's queen, which is seriously damaging, they are not easy or evident in real-life problems, e.g., which R & D project to pursue, not knowing its outcome and impact on the new product decision and not knowing the market acceptance of the new product to be introduced if the R & D effort is success-

5. Hilts, *op cit.*, 230.

ful. Since the new product has no price history, the probability of its success can at best be subjectively assigned. Such subjective assignment is person-specific: as good or bad as the person's judgment. Experiences with similar new products can help; so can careful market research. But they are no substitute for the actual data on the performance of the new product itself which will not be available for months, if not years, *after* the immediate decision to choose or reject a particular R & D project.

0.0 Heuristics — "Serving to Discover"

We need to understand how humans wend their way to decisions through the undefined and often amorphous mass of data and nondata. If the next operation, e.g. developing a new product, depends on the outcome of the preceding operation, e.g., success of the R & D project, neither of which is well-defined, then we need rules of discovery which tell us when we have reached the goal. Creating a path through ambiguities is a process of discovery. Given a group of objectives, not well-ordered (e.g., become famous, make money, do good), and given an alternative set of paths with varying capabilities to achieve the objectives (e.g., go to college, travel abroad, skip school and go to work directly), how does one allocate resources to develop a path to reach the objectives? George Polya used the word heuristic:

> Heuristic, or heuretic, or "ars inveniendi" was the name of a certain branch of study not very clearly circumscribed, belonging to logic, or to philosophy, or to psychology, often outlined, seldom presented in detail, as good as forgotten today. The aim of heuristic is to study the methods and rules of discovery and invention. . . .
> [H]euristic, as an adjective, means "serving to discover."[6]

Unlike in the game of chess, the objectives may not be at all well-defined in real life. In the example of the choice of an R & D project, one objective could be to develop a new or novel method of making a product with certain properties. The failure to develop such a method does not mean that all is lost. When Thomas Alva Edison failed to find in his first 2,814 materials a suitable element for the filament in the electric bulb, his observation was that he found 2,814 materials that did not work. Why a material failed as a suitable filament for the electric bulb was equally, if not more, important than the final success. In other words, the objective itself is changed, refined, and redefined in the process of the experiment or experience. Of course, as the objective changes, so do the succeeding means to meet the objective.

6. George Polya, *How to Solve It* (Princeton, N.J.: Princeton University Press, 1947), 112–13.

Robotics/Artificial Intelligence/Productivity

The process of discovery in human problem-solving is rarely static. To perform better than the previous time, the relationship between objective and means has to be evaluated: how well did the particular means meet the specified objective? Human beings discover *new objectives* however, and allocate new resources to meet the new objectives.

Quality control is the mechanism by which human beings ensure that the given objective is met at least at the specified level most of the time. A machine can make the comparison between the specified performance characteristics and the actual performance of a given product. Since it does not get tired or bored, the quality of the one millionth unit is as dependable as the first unit.

0.0 From the Robotics Revolution to Adaptive Automatics

Now consider the possibility that the machine is not only told to ensure certain performance characteristics, but also to check alternative ways in which the same performance can be achieved. Further, it is provided the comments from which inferences can be drawn on the desired levels of performance characteristics. The assignment is: Specify the performance characteristics which can meet the customer specification at the lowest cost.

We have now moved from the present role of robotics to a potential one, viz., from monitoring performance characteristics to modifying them; from enforcing designed objectives to enlarging them by discovered objectives; and from repetitive production to adaptive production. We define the heart of the Robotics Revolution to be Adaptive Automatics (AA): *the science of incorporating progressive modifications into the automatic process of producing goods and services to better accomplish and/or discovered objectives.*

0.0 Adaptive Automatics (AA) in Molecular Biology

Incorporating progressive modifications into subsequent editions is the hallmark of evolution in nature. The fittest survive because they have learned to live with the environment. That means revising objectives, or means, or both: the long neck of the giraffe being the modification of the means to achieve the objective of reaching the leaves on the tall trees in the environment.

The objectives of each species and each member of each species are written in the counterpart of the human genetic code. If the code can be decoded, it is conceivable to recode it. Since cancer is the multiplication of cells without control, it is crucial to know how the normal cells are coded to be controlled in their cell division.

School children know that a cell multiplies by dividing itself, the billions of precise cells which later grow into the full man or woman growing out of the successful sperm fertilizing the ovum. They also know that their bodies, from the size of their brains to the texture of their toenails, are governed by precise instructions carried by the genes. In 1953, James Watson and Francis Crick announced the

structure of DNA, the material of the genes. Within a year or two of the famous DNA paper, French biologists Jacques Monod and Francois Jacob tried to find the answer to the puzzle created by the fact that if each cell in our body contains all the DNA, then most of the DNA messages must be turned off most of the time, such as the genes for muscle protein existing in toenails. Monod and Jacob suggested that some "repressor" material prevented DNA actions. Their guess was that the carrier of the genetic code, RNA, could be the repressor. But how would RNA or some other chemical turn the genes on and off?

Mark Ptashne has been working on the repressor since 1965 at Harvard. His findings as reported by *Washington Post* writer Philip J. Hilts are as follows:

> Working with Tom Maniatis, Barbara Meyer, and others, [Ptashne] found that the repressor mechanism operates in a *complete loop, like those which form the basis of computer programs. It regulates itself.* The details, in rough terms, are these:
>
> We can picture the [cell] *E. coli*, grazing and gently spinning its motorized flagella. The small lambda [elemental bit of genetic information] settles on the outside of *E. coli*, like a mosquito, and inserts a protein syringe into the *coli*. Through the syringe it injects DNA into the bacterium. The parasite now can either use the *E. coli*'s gene machinery to manufacture more of its kind. *Or it may become dormant.* . . .
>
> When the lambda has become dormant, virtually the *only* one of its fifty *genes that is working* is the *repressor gene*. It must make the repressor to keep the lambda in its sleeping state, and it does this simply by sitting on the "start" signal just ahead of all the lambda genes. . . .
>
> Repressing the repressor is important in switching the system from one stable state—producing repressor and holding back lambda—to the next state—producing lambda and holding back the repressor. In fact, the double-repressor mechanics of the system work so that the change of state occurs quite suddenly. . . .
>
> Many different substances can trigger the SOS response and cause lambda to rise from its dormancy. As it happens, all these substances cause cancer. . . . The substances which cause the lambda to awake and lyse the bacteria are probably carcinogens. The test is about ninety-nine percent accurate.
>
> This result has led Ptashne and his colleagues to think it

possible that there is some fundamental *biological similarity* between cancer and lambda release.[7] (emphasis added)

Ptashne's findings tell us that the repressor mechanism is *self-regulating*. Self-regulation presupposes several things: (i) prespecified objective(s); (ii) investment of resources to fulfill (i); (iii) measurement of results; (iv) comparison of actual results with prespecified results; (v) choice of prespecified alternative course of action depending on comparison of actual results.

Three results are possible under the fourth presupposition. The difference between actual and prespecified or ideal is zero, positive, or negative. If actual-ideal is zero, the system continues without change. If positive, probably reduce the investment of resources. If negative, probably increase the investment of resources.

If dormancy is the ideal state of the system, carcinogens awaken the lambda. This could mean that the actual-ideal is positive. The healthy reaction would be to reduce the investment of the resources to maintain dormancy. Instead, the carcinogens cause an increase, something like adding fuel to a fire.

This possible "biological similarity" between cancer and lambda release also tells us that in healthy cells, the choices of the prespecified alternative course are well-defined and executed. According to Ptashne, who has spent sixteen years studying the lambda, it appears almost as if the lambda virus detects an insult to the cell and so breaks out to evade the danger. When the *E. coli* begins to be damaged, the cell releases an enzyme that cleaves all sorts of repressors in the DNA, one of which, Rec A, cleaves the lambda repressor. The repressor can no longer hold back the "start" signal, and the enzymes begin reading the lambda DNA and carrying out their instructions.

We could look upon the several actors in the molecular biological drama of self-regulation as nature's adaptive automatics. We see an original objective, O_1, of writing DNA instructions to be carried out, modified into a second objective, O_2, of not reading the written DNA instructions, so that they may *not* be carried out at Time T_1. The means, M_1, of recording the DNA instructions are DNA; means, M_2, of blocking the reading of DNA instructions are lambda, and Means, M_3, the second repressor.

When the cell begins to be damaged, a third objective, O_3, may be identified: save the cell. To achieve this objective, the written instructions should be read, so that they may be carried out in Time T_2. The means to achieve this is the unblocking of the reading by Means M_4, the cell releasing an enzyme to cleave the repressors, including the lambda repressor.

The modification of objectives and the modification of the means to achieve them, in the light of experience, is the essence of adaptive automatics, so well-observed in *E. coli*.

7. Hilts, *op. cit.*, 188, 189, 191.

0.0 The Significance of Adaptive Automatics to Technology and Trade

When man imitates nature, as in the case of the airplane imitating the bird, the imitation may look considerably different from the original. When man imitates the objectives and means of modification found in the cell *E. coli*, the imitation is also likely to be vastly different from the original. Notice that we define adaptive automatics as the science of incorporating progressive modifications into the automatic (instead of the organic) process of producing goods and services (instead of preserving and multiplying organisms) to better accomplish designed and/or discovered objectives.

Adaptive automatics, as well as its subsets of the robotics revolution and artificial intelligence, holds enormous significance to the international flow of technology and trade.

0.0 Industrial Robots, the Beginning

An industrial robot is essentially an arm which can move in many directions. Advanced technology increases the level of sophistication in the field. For instance, adding vision to the robot permits its use in inspecting products for quality control. The payback period of capital investment has been falling, making the introduction of robotics economically feasible.

Robotization is no respecter of boundaries. Properly installed and operated, the robot will perform as well in Korea or Japan as it does in the USA. Robots do not require that the nation shall have gone through all the various stages of development that Japan or the USA went through.

Several developing nations, such as Korea, Taiwan, and Hong Kong, can *leapfrog into robotics*.

The question is: Does Korea (Taiwan, Hong Kong) *want* to leapfrog into robotics? Does Korea (Taiwan, Hong Kong) *need* to leapfrog into robotics?

Robotics today is perhaps where computers were in the 1950s. It will be recalled that repetitive calculations were done by physically wiring boards which were inserted into the computer of the day. With every change in the process of calculation, new wiring had to be made. While wiring boards will not even be recognized by any but the senior citizens of the computer community, they did represent a major breakthrough from the typewriter-size calculators which were operated by literally cranking the numbers up manually. An entrepreneur cornering the wiringboard market would soon be left high and dry when programming languages appeared by the latter part of the 1950s; and certainly when the First Generation computers were replaced by the Second, and, by the mid-1960s, the Second by the Third.

0.0 First-Generation ADAPTOMATONs

Thirty years after the days of the wiringboard, Japan is vigorously planning to launch the 5G—the Fifth Generation Computer. When it emerges, the focus will

be not on number crunching, but on symbolic manipulation; not on pernicious precision in programming, but on reasonably coherent logical steps, even with some ambiguities; above all, not on numerical answers carried out to the nth decimal place, but on *knowledge-processing*. Excessive investments in one-armed metallic monsters may appear ungainly by what we may call First Generation ADAPTOMATON, embedding ADAPTive autOMATics. The ADAPTOMATON would differ from the robot primarily in its capacity to discover and modify objectives and accomplish them better, and to modify the means, i.e., develop goals, and design methods (DGDM). The ADAPT part of the ADAPTOMATON *adapts* the goals to the environment, upgrading them, as it were, based on meta-goals either prescribed or discovered. The OMAT part from Automatics *designs* the methods to meet the goals, the ON-ending of the name suggesting *automaton: An ADAPTOMATON is a self-discovering and self-designing system which transforms inputs into outputs progressively and efficiently to accomplish designed and/or discovered objectives.*

While the world waits for the First Generation ADAPTOMATON, what should the Newly Industrializing Countries (NICs) do with respect to robotics?

0.0 PSPG, Ratio of Performance-Specific/Performance-General

*Since the import of high-technology, such as robotics, by NICs is to export high-*tech products back to the high-technology exporters themselves, the demand for high-tech products should be a critical concern. The US and Japan are prime sources of high-technology for NICs and for the rest of the world. What do the US and Japan look for by way of high-tech products?

Today's list of preferred high-tech products may not remain the same tomorrow. Therefore, more stable criteria should be applied to develop a strategy of importing high-technology from the US and Japan in order to export to them high-tech products and services.

The issue of technology imports today to increase export earnings tomorrow will *not* be resolved in one-, two-, or three-year time-periods. It should be seen, instead, in the context of a shifting cycle of functional specialization. In the mid-1960s, Korea, Hong Kong, and Taiwan concentrated on clothing, footwear, toys, and basic electronic assemblies, which required little capital or technology, but which were labor-intensive. Ten years later, they moved into capital-intensive processing industries. Korea, for instance, has the world's largest shipyard. Construction and manufacturing sectors, with their respective gains in the first quarter of 1983 of 23.6 percent and 10.7 percent over the first quarter of 1982, contributed most to the 9.3 percent surge in Korea's GNP.[8]

8. *Business Korea*, Vol. 1, No. 2, August 1983.

CONCOLs in Trade and Technology

To select a technology strategy for Korea, we have to consider the already-observed cycle of specialization:

- 1965: low capital; low technology; low-skilled labor; simple products
- 1975: high capital; medium technology; high-skilled labor; complex products
- 1985: high capital; high technology; higher-skilled labor; more complex products?

In the 1960s and 1970s, Japan went through the phases now experienced by Korea. What Japan has done is to move into products and processes which require a special-skilled labor force, such as fiber-optics, large-scale integrated circuits, sensor devices, advanced aircraft engines. Japan has also committed itself heavily to robotics in industry.

What type of high-tech products is Japan interested in; what type of products is the US interested in?

Korea's exports in 1982 were $21 billion. Several systems, each costing much more than $22 billion, are currently being developed in the US. In fact, the expenditures on the 182 Major Systems (each of which costs $500 million in production or $75 million in research and development) in fiscal year 1984 (1 October 1983–30 September 1984) was $486 billion. For every dollar in acquisition cost, these systems require between $2 and $3 in operating costs, making the total expenditures for acquisition and operation range between $1,448 and $1,944 billion, or $1.5 to $2 *trillion*. In other words, 182 Major Systems cost over twenty-five to thirty years, half to two-thirds of the entire GNP for a year!

Each of the 182 Major Systems is *one-of-a-kind*. Several units of each system are produced, but each production run is sufficiently different from every other production run so as to make each system perform very specific functions (in terms of accuracy, throwpower, etc.).

The long leadtime in developing each weapon system is devoted to discovering modified objectives and designing the weapon system to meet them, a process uniquely suited to ADAPTOMATONs. If the First Generation ADAPTOMATONs were available, they surely would cut down the leadtime in design, development, and production of the 182 Major Systems.

We notice that these systems are each required to perform specific functions, or they are *performance-specific*. Contrasted with performance-specific items are the products off an assembly line, any one of which would perform just as well as any other, so that they are called *performance-general*.

The trade strategy question for Korea is: How much of Korean exports in the 1990s should be performance-specific and how much should be performance-general? If PSPG, which is the ratio of performance-specific/performance-general, is an accepted criterion, its present state may be, say, 0.5 percent and its potential state in 1990 may be, say, 47 percent.

0.0 The Concept of Concomitant Coalitions (CONCOLs)

The fact that Korea and other NICs want to import high technology from the US and Japan today to export *high-tech products* from Korea to the US and Japan tomorrow points up an inherent conflict.

0.0 Inherent Conflict in Trade

Exports by Korea are imports for other countries, Korea earning what the importing country "loses." The "loss" sustained by the importing country is foreign exchange, balanced by the "gain" for the importing country of goods and services received from Korea. Therefore, when exports, "the engine of growth" of Korea, advance powerfully, that advance must be paid for by other countries. To pay for increasing imports, the importing countries must earn foreign exchange from Korea, from developing countries, and from advanced countries. The more successful Korea becomes in export competition, the more successful its customers must become in earning (or borrowing) foreign exchange.

0.0 Competitive Difficulties on Both Sides

President Chun Doo Hwan, of the Republic of Korea, summarized the current position of Korea in the exports competition: "Korean exports have almost reached their limit, currently undergoing difficulties in competition with less developed countries for the lower end of the markets as well as with advanced nations for the upper end of the markets."[9]

The "upper end" usually implies high-technology items. Here, the paradox of the relationship between the importer of technology (Korea) and the importer of goods from Korea (USA, Japan) emerges vividly. To enhance exports, Korea must import technology; to pay for technology imports, Korea must export more. (A Catch 22!). We can present this inherent conflict in international trade relations as follows:

Importing high technology from the US and Japan in order to export high-tech products to them places Korea *simultaneously* in opposing roles:

- Korea *with* the US (paying more for high-tech)
- Korea *against* the US (demanding more for high-tech products exported)

This association of one party with and against others in the same game is called concomitant coalition (CONCOL).

0.0 Discrete Coalitions

In *Theory of Games and Economic Behavior*, John von Neumann and Oskar Morgenstern discuss three-person games in which player A can join *with* player B, *ipso facto* aligning *against* player C, to pursue maximization of returns to the coalition of A and B at the expense of C. A coalition is an association of opposing in-

---

9. *Business Korea*, August 1983, 25.

terests. In the three-person game, which coalition—AB, AC, or BC—will be formed will depend on two things: (i) the probability of 0.5 or more that the coalition outcome will be greater than the sum of the coalition member outcomes, and (ii) the prior agreement on the distribution of the coalition outcome among coalition members, known as the *imputation*.

Equally, if not more, important than the coalitions themselves is the *bargaining* that precedes their formation or nonformation and/or dissolution. Bargaining is the process of offer and counteroffer of future outcome(s) in return for present association(s).

The three players A, B, and C are in conflict with each other. A is against B in that both A and B are trying to maximize their individual outcomes. When the total outcome is fixed, A's gain can be only at B's expense, and vice versa. *Conflict is the commitment of competing inputs for identical outcome.*

In the Neumann-Morgenstern formulation, the coalitions are exclusive and exhaustive:

- A with B against C
- A with C against B
- B with C against A

0.0 Concomitant Coalitions in Nuclear Survival

In real-life, coalitions may not necessarily be exclusive and exhaustive. Consider alliances between nations. Take, for instance, the nuclear survival "game." The United States and France are both members of the North Atlantic Treaty Organization (NATO) and the USSR is not. We can present this alignment of opposing interests as:

US *with* France *against* USSR on NATO

The first significant treaty to lessen the possibility of nuclear war was the Test Ban Treaty, which was signed by the US, USSR and most other nations, but not by France. The "game" is still the same, viz., nuclear survival, in which the interests are:

US *against* France *with* USSR on Test Ban

Both NATO and the Test Ban Treaty are integral elements of the same "game" of nuclear survival. We find that the same nations play with and against the same nations in the same game. They simultaneously compete and collaborate, forming opposing coalitions. This new concept in game theory was defined in 1961:

Concomitant coalitions (CONCOLs) are distinguished by the following features:

1. Simultaneous entry into coalitions by players,
2. who band together for and against the same party at the same time, and

3. make moves subject to the several objectives of the different coalitions, but
4. governed ultimately by the overriding principle of self-interest,
5. in order to achieve joint profit-maximization,
6. in a nonzero-sum game.[10]

We can define concomitant coalition as the association of one or more parties with and against other parties in the same game.

The parties can even be the same person:

> Whether we consider love or light, inherent in love is hate, and inherent in light is shadow. Yet these traces of the opposites are discernible only under analysis or in the case of breakdowns, when the tension of the opposites becomes too devastating to be controlled by the organism.
>
> These opposites are a fact of life. There are no "absolute" attitudes. Thus, statements of the form: A "absolutely" loves B, or B "absolutely" hates C, are at best inaccurate; at worst, a horrible camouflage for half-understood notions. At the height of A's love for B, there is *simultaneously* present A's hate for B: only it is not perceptible, just like stars at midday. . . .
>
> Perhaps, this opposition may be viewed constructively. In the words of Charles Dickens, "We are two halves of a pair of scissors, when apart, Pecksniff, but together we are something." While the organism's alter ego seeking to annihilate, and the organism's anti-ego seeking to accommodate would when apart be ruinous, but together would be creatively constructive.[11]

0.0 Concomitant Coalitions in Arms Control

The opposing interests are imbued with personality when reference is made to alter ego and antiego. We find this particularly useful in understanding the concept of harmony, which lies at the very foundation of Japanese life. An example of alter and antiego terminology applied to a real-life problem follows:

10. George K. Chacko, "Bargaining Strategy in a Production and Distribution Problem," *Operations Research*, Vol. 9, No. 6, Nov.–Dec., 1961, 817.
11. George K. Chacko, *Today's Information for Tomorrow's Products—An Operations Research Approach*, (Washington, D.C.: Thompson, 1966), 185–87.

They [the alter ego and anti-ego] inhabit the same organism (the individual, the community, the nation); therefore, that occupancy should be emphasized. They are called alter ego and anti-ego to signify that they represent opposing interests while simultaneously inhabiting the same organism. An example of such inherent opposition of interests is provided by the approach of super powers to arms control. The negotiations are intended neither to abolish wars nor to abolish weapons, but only to reduce the probability of war by controlling the production and employment of weapons. Notice that if a nation agrees to curtail its arsenal, that would be militarily against the concept of providing security to the nation by means of weapons. However, to the extent that the reduction of the increase in weaponry also reduces the possibility of the use of these weapons, to that extent it furthers the aim of the country to survive. The alter ego of the nation would opt for the reduction of the increase in weaponry, while the anti ego would opt for the increase in weaponry. The position of the alter ego is really one of collaboration with the enemy. In any discussions of arms control, "there is also an element of collaboration with the adversary involved."[12]

0.0 Time Horizon in CONCOL

Consider the export trade "game." Korea is *with* the US in technology imports in increasing US export earnings. Korea is *against* the US in high-tech exports in increasing US export payments.

We are implicitly assuming that the US wants to increase export earnings and decrease export payments. When Korea furthers US interests, it is seen as being with the US. If we assume that Korea is likewise interested, we have the following CONCOL interests:

- Korea is *against* the US in technology imports in increasing Korean export payments.
- Korea is *with* the US in technology imports in increasing Korean high-tech export earnings.

Of course, US technology imported today will *not* increase Korean high-tech exports to the US today. Therefore, the last CONCOL should be modified to read:

- Korea is *against* the US in technology imports in increasing Korean export payments *today*.

12. George K. Chacko, *Applied Operations Research/Systems Analysis in Hierarchical Decision-Making, Vol. II* (Amsterdam, The Netherlands-North-Holland, 1976). 322.

- Korea is *with* the US in technology imports in increasing Korean high-tech export earnings *tomorrow*.

Time is a critical dimension in CONCOL. Those who argue for technology imports from the US today are taking the perspective of tomorrow in justifying the increased export payments today. On the other hand, those who argue against increased technology imports from the US today are justifying it in terms of not increasing Korea's export payments today.

While, from the point of view of conserving foreign exchange today, limiting US technology imports would be sound, President Chun's reference to the competition with advanced nations for the upper end of the markets appears to consider the long-term capability to compete, which definitely requires technology that advanced nations have to offer. When opposing time-horizons enter into the discussion with reference to the same party (country), it is convenient to speak of alter ego and anti-ego. For instance, alter ego could argue for the long-term whereas anti-ego could argue for the short-term. We could personify technology viewpoints in Korea as the following CONCOL:

- Korean alter ego is *with* the US in technology imports in increasing Korean high-tech export earnings *tomorrow*.
- Korean antiego is *against* the US in technology imports in increasing Korean exports payments *today*.

The roles of alter ego and antiego are quite realistic in US corporations. The Chief Executive Officer (CEO) is supposed to be so fully preoccupied with the long-term that he would hardly know what goes on in the short-term. On the other hand, the vice president of Production would be so immersed in the short-term that he would be barely oriented to the long-term. To designate their respective functional roles as alter ego and antiego is not to imply that either is less committed to the welfare of the corporation: simply, that each is expected to advocate vigorously his assigned role and perspective.

It is not as though the alter ego or the anti-ego has to dominate. As a matter of fact, sometimes the alter ego considerations overrule the antiego considerations, and vice versa. Just as in the corporation, so also in the nation: sometimes, the alterego considerations are dominant; sometimes the antiego considerations are dominant.

0.0 CONCOL Between the US and Japan

Importing high technology is by no means an either-or proposition, as shown by the recent General Motors-Toyota agreement. One specific instance is the reopening of the California plant of General Motors, which had been closed down. Why would Japan want to give employment to Americans in California, employing Japanese technology? The answer is that Toyota wants to produce a specific type of small car in California. Toyota would bring about $1 billion (US) in advanced machinery, while General Motors would bring about $1.25 billion (US)

in plant and land to the CONCOL to capture a specific segment of the auto-market at a given time.

0.0 Market Segmentation

The answer to the question, How much of Korea's exports should be performance-specific in the 1990s and how much should be performance-general? addresses both type of difficulties referred to by President Chun Doo Hwan. In general, the performance-general products and services will find export markets in the developing countries and the performance-specific products and services will find export markets in the advanced countries. The key is choosing a strategic segment of the market for the given product ideal for the nation, and capturing that segment, not for all time to come, but for specific predetermined periods.

CONCOL provides the philosophical framework to formulate Korea's problems with the lower end of the markets as well as the upper end of the markets. In applying CONCOL to practical problems, it has been found necessary to state *operationally* the goal(s) and subgoal(s) of the party or parties concerned. Identifying the CONCOL interests is the first step. Next, the numerical description of the present state, and the potential state. For instance, the parties to Korea's export trade will include different ministries assigned administrative responsibilities, major corporations engaged in export, as well as associations of export interests. If PSPG—the ratio of performance-specific/performance-general—is an accepted criterion, its present state may be, say, 0.5 percent and its potential state in 1990 may be, say, 47 percent. How to move from 0.5 percent to 47 percent will be the concern of the *process* of CONCOL.

9

CONCOL Players Promoting Productivity in Japan

OVERVIEW: The institutional, investmental, and inventional/innovational factors of Japanese productivity can be considered from the point of view of Concomitant Coalitions (CONCOLs) in which the same party plays with and against the other party(ies) in the same game.

When the opposition is within the same player, the terms "alter ego" and "anti-ego" serve to personify those opposing interests. Thus, the company alter ego can be presented as playing *with* long-term planning *against* short-term profits; while, the company anti-ego plays *against* long-term planning *with* short-term profits. The personification explicates the inherent nature of the conflict and underscores the fact that the opposition resides within the same entity, i.e., the company.

The CONCOL players are identified in eight institutional, nine investmental, and seven inventional/innovational elements. The opposing interests are reconciled or harmonized through the CONCOL bargaining process, which will be discussed in chapter 11. In the meantime, the twenty-four major CONCOLs identify the significant opposing interests whose skillful balancing accounts for the visible success of Japanese productivity. To emulate this, the first step is to identify similar sets of CONCOL players in the American economy.

217

0.0 CONCOL in the Institutional Factors
of Japanese Productivity

The thesis of this chapter is that the American concept of Concomitant Coalitions (CONCOLs) helps us understand the triad of Japanese productivity, permitting practical applications with and between both the countries.

In chapter 3, we identified what we called the triad of Japanese productivity: institutional factors, investmental factors, and inventional/innovational factors. We will now interpret the first one in terms of Concomitant Coalitions.

As we pointed out in chapter 3, the concrete illustrations were developed through quotations, so that in this chapter, we merely summarize the argument, inviting the reader to the original source materials already presented earlier.

0.0 Long-term Plan, Not Short-term Profit

Shibusawa, whom Frank Gibney calls the "conscience of Japanese business," made an effective plea to plan wisely for the future of the company and the country. Clearly, no business could long endure without receiving profits. What he emphasizes is that profit will accrue in the long run to those who planned well. We may state this principle in terms of CONCOL interests, personified as players, as follows:

- Company alter ego *with* long-term planning *against* short-term profits *with* company growth

- Company anti-ego *against* long-term planning *with* short-term profits *with* company survival.

While both planning and profits are integral to the survival and growth of the company, the *primacy* given to planning would, to that extent, deny the short-term profits, and vice versa. This conflict of primacy is personified in the alter egos and anti-egos.

Matsushita, the founder of the largest electrical appliances manufacturing company in the world, echoes the view that profit is something that accrues to the deserving company:

> A business should quickly stand on its own, based on the service it provides the society. Profits should not be a reflection of corporate greed but a *vote of confidence* from society that what is offered by the firm is valued.[1] (emphasis added)

He recognizes that the vote of confidence takes time; it is not won overnight. Matsushita expects every division to be self-sufficient within five years.[2,3] The

---

1. A. Takahashi and H. Ishida, "The Matsushita Electric Industrial Co. Ltd. Management Control System," Business School, Keio University, 1971, 111.
2. *Ibid.*, 10ff.
3. Rowland Gould, *The Matsushita Phenomenon*, Diamond Sha, Tokyo, 1970, 55 ff.

dimension is time; the competing variables are service and profit. What Matsu-shita holds is that if the division concentrates on service, it will win the confi-dence of the society it seeks to serve, and profits will accrue. On the other hand, if the division concentrates on profit, probably at the cost of service, it will not win the confidence of the society it seeks to profit from, and profits will not ac-crue in the long run. In general, the concomitant coalition interests (players) are:

- Company alter ego *with* long-term service *against* short-term profits *with* company growth
- Company anti-ego *against* long-term service *with* short-term profits *with* company survival.

0.0 Employment of Men and Women in the Principal Firm
Long-term service to the customer is provided by the *lifelong* employee of the company. There are several critical elements in the lifelong employment practice. *The first critical element is the protection of lifetime employment at the expense of temporary employees.*

Ouchi points out that women in major Japanese firms are considered tempo-rary employees, even though they may work for 20 years. As employees, they are counted upon to provide the service that makes the company grow in the long-term, but they are the dispensable part of the work force. We can then present the following concomitant coalition players:

- Company alter ego *against* long-term employment of women
- Company anti-ego *with* long-term employment of women.

This sacrifice of the long-term employment of women is made to protect the long-term employment of men:

- Company alter ego *with* long-term employment of men *against* long-term employment of women
- Company anti-ego *against* long-term employment of men *with* long-term employment of women.

0.0 Employment in the Satellite Firms
The second critical element in the lifelong employment practice is the protection of lifetime employment of male employees in the principal firm at the expense of both male and female employees in the satellite firms.

The five hundred or more satellite companies which supply specific parts and/or products to Matsushita are the first ones to feel the pinch of business fluc-tuations. Since the satellite firms deliver right to the floor the parts needed to keep the principal firm's production line going, and since the satellite firms produce and store the parts on their premises until the last moment they are called for by the principal firm, in times of business slowdown, the satellite firms are left hold-ing the bag. The principal firm simply turns its demand off, leaving the satellite firms to fend for themselves. Precisely the parts, the products, and the services

which are most susceptible to fluctuations are the ones that the principal firm contracts out to the satellites. If the principal catches a cold, that means pneumonia to the satellites. And the pneumonia of lower demand can well be fatal to the satellites, presenting the following concomitant coalition players:

- Principal firm alter ego *against* long-term demand (stability) for satellite inputs
- Principal firm anti-ego *with* short-term demand (stability) for satellite inputs.

There is another side to the coin. That is the top-down loyalty of the principal firm to the satellite firms. The satellite firms need the principal to obtain licenses to import raw materials necessary to manufacture parts for the principal, and they depend on the principal for their very survival. The life-giving license is closest to the *On* received from one's lord. It entails a *Giri*, repayment with mathematical equivalence to the favor received within the appropriate time limits. While the principal firm will curtail the demand for the goods and services of the satellite firms to protect itself in the short run, the very existence of the satellite firm is based on the long-term commitment by the principal firm, suggesting the following concomitant coalition players:

- Principal firm alter ego *with* long-term commitment to the satellite firm
- Principal firm anti-ego *against* long-term commitment to the satellite firm.

We can relate the intended consequence to the long-term employment of men at the principal firm:

- Principal firm alter ego *with* long-term employment of men *against* long-term demand (stability) of satellite inputs *with* long-term commitment to the satellite firm
- Principal firm anti-ego *against* long-term employment of men *with* short-term demand (stability) for satellite inputs *against* long-term commitment to the satellite firm.

0.0 Lower Weekly (Monthly) Remuneration of Men and Women in the Principal Firm
The third critical element of the lifelong employment practice is the protection of the lifetime employment of male employees in the principal firm at the expense of the weekly paycheck.

If the demand falls off, the reduced revenue has to result in lower total costs; otherwise, the company will have to spend more than it receives. If the demand fluctuates violently, wages and salaries will also fluctuate violently if they are tied directly to the revenue.

What the Japanese firm does is to pay *lower wages*, usually for six months. At the end of the period, everybody receives the same fraction of their salary as bonus. The bonus depends on the performance of the firm, not of the individual.

So, if the demand for the firm's products and services has been good, the bonus will be healthy—possibly equal to half a year's salary. By paying, say, $1,000 each month, instead of $1,500, the firm has a float of $6,000. If business is good, it pays out the $6,000 in two installments: in June and in December. If business is bad, it pays out, say $1,000 in two installments. In other words, the employee continues to get $1,000 every month (plus the $1,000 in bonus), making the total earnings $13,000 for the year even though the times are bad. If times are good, he earns $18,000 for the year. Since he is used to living on $12,000 basic salary, the bonus comes as a pleasant surprise—not the fact, but the size of it. Even when the surprise is not too pleasant, he still keeps his job.

The lower wage (of $1,000 instead of $1,500) suggests the following concomitant coalition players:

- Employee alter ego *with* long-term employment stability *against* higher wages *with* higher bonus fraction
- Employee anti-ego *against* long-term employment stability *with* higher wages *against* higher bonus fraction.

0.0 Remuneration According to Seniority
The fourth critical element of lifelong employment is the protection of the self-image of the lifetime employee in the principal firm through the seniority system.

When one is assured of working for the company for one's entire working life, he can accept $1,000 a month instead of $1,500, knowing that, unless times are really bad, he will receive $6,000 in bonuses. How is the $1,000 determined?

Ouchi narrates the anecdote of a Japanese-owned, American-operated electronics factory which tried a piece-rate system: the more you wired, the more you got paid. Two months after the opening of the factory, during which time it was operated under the incentive system, the plant manager was faced with en masse resignation at the end of the week. Reason? The incentive system.

The foreman advised the plant manager that no one could be more productive than anybody else, because no one in the final assembly phase could do a thing without everybody else having done her job. Higher pay for larger wirings, then, would single out the final assembly person, even though she could not have done what she did without the backing of everyone else. Those on the front end of the assembly-line were not recognized in the piece-rate incentive system. How should they be paid? By seniority:

> When you hire a new girl, her starting wage should be
> fixed by her age. An eighteen-year-old should be paid
> more than a sixteen-year-old. Every year on her birthday,
> she should receive an automatic increase in pay.[4]

4. Ouchi, *op. cit.*, 41.

Robotics/Artificial Intelligence/Productivity

The principle of seniority is arrived at on the basis of group productivity, as opposed to individual productivity. While the foreman did not explicitly state it, the idea is that an eighteen-year-old should be paid more than a sixteen-year-old because the former has worked at the job for two more years, presumably gaining skills and experience which contribute more to her productivity than when she started on the job. This suggests the following concomitant coalition players:

- Employee alter ego *with* group productivity *against* individual productivity *with* seniority system *against* incentive system
- Employee anti-ego *against* group productivity *with* individual productivity *against* seniority system *with* incentive system.

0.0 Generalist in the Company, Not Specialist in the Profession
The fifth critical element of lifelong employment is the intensive concentration within the company at the expense of professional specialization.

Parallel to the eighteen-year-old being automatically given a pay raise on every birthday is the practice in the professional ranks of being given exactly the same increase in pay and exactly the same promotions as one's peers entering the service of the company at the same time.

With one exception. At the end of ten years, one is evaluated seriously to determine if one merits a larger promotion and pay raise than another. The presumption is that ten years is a long enough time to demonstrate one's potential for productivity.

For being allowed to learn all about a company so that one can work in any and every part of it, an employee pays a price—he or she can work only for that company. As Ouchi puts it:

> In the United States we conduct our careers between organizations but within a single speciality. In Japan people conduct careers between specialties but within a single organization. . . . In the United States, companies specialize their jobs and individuals specialize their careers. . . . Japanese do not specialize only in a technical field; they also *specialize in an organization*, in learning how to make a specific, unique business operate as well as it possibly can.[5] (emphasis added)

Because a Japanese employee specializes in a particular organization, he or she is company-specific. Even in the same industry, each firm has its own unique characteristics, so that an excellent man in Nuts and Bolts Company A is out of his element in Nuts and Bolts Company B. Gibney explains why:

5. Ouchi, *op. cit.*, 29.

222

Living in the storied land of interchangeable employee parts, we tend to think that anyone who can do his job well should fit into almost any company. It is largely a matter of qualifications. Not so in Japan. Dai Nippon Manufacturing Company K.K. may be in the same kind of business as Teikoku Nuts and Bolts K.K., but their people may be quite different. Where Dai Nippon's executives are virtually an old-boy network of Keio graduates, Teikoku's people are recruited from Waseda. Where one has a long relationship with the Mitsui Bank, the other may rely on Suminoto. Where one does most of its production in-house in its own plants, the other may rely on a network of dependent subcontractors and subsidiaries. One favors tapping high-ranking bureaucrats to join its board of directors, the other does not. Such differences matter. They represent not merely a different choice of serving but a different set of attitudes, in which two different sets of people have grown up. In short, even where companies perform the same functions, they will have different identities.[6]

We may present the CONCOL players as follows:

- Employee alter ego *with* organizational specialization *against* professional specialization *against* occupational mobility
- Employee anti-ego *against* organizational specialization *with* professional specialization *with* occupational mobility.

0.0 Continuous Training by the Company
The sixth critical element of lifelong employment is the ongoing training of the individual by the company, an effort that would be lost to competition if there were widespread occupational mobility:

One effect of lifelong job rotation is employer incentives to develop the skills and the commitment of their employees. In the United States employers are reluctant to invest in training employees in new technical skills, because a skilled employee can easily find another employer, taking along that investment.[7]

Militating against such a loss of the investment in the technical training of the employee in the Japanese company are two factors: the lifetime employment com-

6. Gibney, *op. cit.*, 77.
7. Ouchi, *op. cit.*, 29.

mitment and nontransferable company-specific skills, excellent for only the one particular company:

- Company alter ego *with* continuous employee training *with* long-term productivity *against* occupational mobility
- Company anti-ego *against* continuous employee training *against* short-term costs *for* occupational mobility.

0.0 Continuous Acculturation by the Company
The seventh critical element is continuous training in the values of the company.

The continuous technical training is one side of the coin; the continuous cultural training is the other. At Matsushita, which was the first company to introduce the company song in Japan, there are two distinct kinds of training provided. One is basic skills in training; the other, more fundamental one is training in the Matsushita values.[8] These values are acquired by self-indoctrination.

Again, when one gets self-indoctrinated into the values of a particular company, it is a commitment for life. Above all, it makes it possible for company management to train its employees in new technical skills without having to look nervously over their shoulder to see if the employees are being wooed away by the competition:

- Company alter ego *with* acculturation *with* long-term commitment to company *with* long-term commitment of company
- Company anti-ego *against* acculturation *against* long-term commitment to company against long term commitment of the company.

0.0 Summary of Institutional Sets of CONCOL Players
The fundamental characteristic of the institutional factors of Japanese productivity is the primacy of the long-term.

While long-term planning, well-conceived and carried out, is most likely to assure customers of the quality and care of the products and services offered by the company, thus leading to ever-expanding sales and profits, it cannot be done without any thought to short-term profits. To the extent that the long-term planning overrules the short-term profits, the two are in conflict. *Both* planning and profit, however, are integral to the survival and growth of the company. Since both exist simultaneously, they may be personified in the company alter ego and company anti-ego, the alter ego aligning *with* the long-term planning *against* short-term profits, the anti-ego, the reverse.

When long-term planning is given primacy, it sets the stage for lifetime employment; for it is the people who have to plan, implement the plans, and make the profits by rendering products and services valued by the customer. To ensure

8. Tohiko Nimura, "Characteristics of Personnel Management at Matsushita," in *Business Behavior in the Home Appliance Industry in Japan*, ed. Yasuo Okamoto, Monograph (Tokyo: Tokyo University Economic Research Fund, 1973), 148–66.

that the men have lifetime employment, women are treated as temporary employees, no matter how long they serve the same company. Further, to protect the lifetime employment of the men of the principal firm, the goods and services subject to most fluctuations are contracted out to satellite firms, which have the principal firm as the sole customer. The dips in demand for the principal firm's products and services translate immediately into decreased demand for the satellite firm's products and services, leading to layoffs and even bankruptcies. Again, to protect lifetime employment, the workers are paid lower weekly (or monthly) wages, and paid bonuses equalling as much as 6 months pay in two (or three) lump sums. Even if the lump sum payments are slim or occasionally have to be skipped altogether, the company is able to keep the men on the payroll at regular wages, none having to be laid off even if the company experiences a 30 percent decrease in revenue.

To permit the lifetime employment with the same company, the remuneration is based primarily on seniority, at least for the first ten years of service. When one is committed to serving the company all one's life, he is relieved of the pressure for immediate performance and allowed to grow with the company. Each person is trained in the various operations of the company, so that the specialization is company-specific. The company invests heavily and continuously in training employees in new skills, on the one hand, and in self-indoctrination of the company values by the employees, on the other.

The value of CONCOL is in explicating the principal perspectives of the survival and growth of the company that are in inherent conflict with each other. Our identification of the personified alter ego and anti-ego opposition illustrates the type of opposing interests on which the chief executive officer has to give his ruling. Whether he gives the ruling directly, which is rare in Japan, or indirectly, the very explication of the opposing aspects of the same goal of *profitable long-term performance of the company in service to the community* enables the emphasis upon one or the other at any given moment.

0.0 CONCOL in the Investmental Factors of Japanese Productivity

Among the investmental factors that account for Japanese productivity, the most significant by far is the role of MITI. This role may be studied in terms of its several manifestations.

0.0 MITI's Promotion of Specific Industries

To MITI's selection of specific industries, Japanese industry and business owe much; for without such a specific focus, comparable results would not be imaginable. In the 1950s and 1960s, heavy industries and chemical industries were the focus of development. In the 1970s and 1980s, it has been knowledge-intensive industries. The 1950s slogan "Steel is the Nation" succeeded, because Japan pro-

duces high-quality steel at the lowest cost in the world, enabling Japan to supply 50 percent of global tonnage construction, 30 percent of global auto production, and a significant percentage of home appliances, machine tools, etc.

You cannot emphasize one without de-emphasizing something else. To emphasize all is to emphasize none. When MITI emphasizes Very Large Scale Integration (VLSI), therefore, it is de-emphasizing some other contenders. In a CONCOL, these contenders need to be identified:

- Country (MITI) alter ego *with* VLSI *against* heavy industries
- Country (MITI) anti-ego *against* VLSI *with* heavy industries.

Chronologically, the most recent specific industry sponsored by MITI was computers. So, we could say:

- Country (MITI) alter ego *with* VLSI *against* computers.

One purpose of sponsoring VLSI activity, however, is to make microcomputers. What MITI is giving up, then, is the sponsorship of computers as a growth industry. Since the computer industry has already established itself in Japan, MITI no longer needs to sponsor it. If microcomputers are a prime object of VLSI:

- Country (MITI) alter ego *with* VLSI *against* miniframe computers *with* microcomputers.

0.0 MITI's Adjustment of Supply and Demand

Even when steel is eminently successful as an industry, MITI carefully identifies the projected needs of big steel users and issues a quarterly "Outlook for Steel Supply and Demand," so that steelmakers can plan their production.

- Country (MITI) alter ego *with* steel users *against* steel producers in projected steel demand
- Country (MITI) anti-ego *against* steel users *with* steel producers in projected steel supply.

The inherent conflict between the supply and demand is in that the former has to serve the latter if MITI, by its documentation, states what the demand is, with the implicit requirement that the supply better meet the demand. If, on the other hand, MITI canvassed all the suppliers and published the outlook for *supply* with the implicit requirement that the demand better not exceed it, the suppliers would have the upper hand. Such an emphasis is indeed preceivable in MITI licensing imports of foreign technology and in fighting with foreign governments on the export of Japanese products to foreign countries. For instance, the negotiations with the United States of the Japanese automobile export quota proceed on the premise that everyone knows what the supply is, with the clearly implied requirement that the demand worldwide better adjust to it. Since the Japanese supply of automobiles must be sold, the demand for it in the United States and elsewhere should do the adjusting.

CONCOL Players Promoting Productivity in Japan

0.0 MITI's Promotion of Production Abroad

While MITI has been stoutly defending the right of Japanese products to access to the world market, it has also been quietly persuading the Japanese manufacturers to start automobile (and other) factories in the United States. To the extent that production is carried out abroad, employment is diminished at home (or at least, the possibility of expansion is reduced), and capital flows out abroad. MITI's urging this expansion abroad, however, is not pure altruism, but a calculated bid to preserve and expand Japanese export markets. Says Gibney:

> MITI, for example, now calls Japanese carmakers in for administrative guidance about the number of cars they should export. It is *MITI* that for some time has been *pressuring* Toyota, Nissan, and others, *to put up plants in the United States.*[9] (emphasis added)

The CONCOL players are:

- Country (MITI) alter ego *with* long-term export market *with* Japanese production abroad
- Country (MITI) anti-ego *against* long-term export market *against* Japanese production abroad.

0.0 MITI's Promotion of Bigness

The MITI policy of identifying growing industries—or rather, industries which ought to be growing, such as heavy and chemical industries in the 1950s and 1960s, knowledge industries in the 1970s and 1980s—also leads them to favor efficiencies of production obtained from control over larger resources through mergers. This brings them into conflict with another agency of the government—the Fair Trade Commission (FTC)—giving rise to the following concomitant coalition interests, personified in players:

- Country (MITI) alter ego *with* rationalization *with* mergers *against* free enterprise
- Country (FTC) anti-ego *against* rationalization *against* mergers *with* free enterprise.

MITI has managed an impressive record of rationalization through mergers despite the opposition from the FTC, Ministry of Finance (understandable territorial struggles), and industrialists and businessmen suspicious of redtape:

> In 1964, ninety-seven shipping companies were consolidated into six groupings. A five-year moratorium on in-

9. Gibney, *op. cit.*, 137.

terest payments to the Japanese Development Bank was offered as an inducement to the industry that was less than enthusiastic at the beginning. Also in 1964, the three former Mitsubishi Heavy Industries firms merged to form a new Mitsubishi Heavy Industries. In 1965, a merger was consummated between Nissan Motor (the second largest automobile maker) and Prince Motor (the fourth largest) after much arm-twisting by the MITI.[10]

0.0 MITI's Placement of Retirees in Commercial Bank

MITI's pervasive influence in Japanese industry and business comes not only from its regulatory powers, which help make it perhaps the most important arm of the government, but also from its *exclusive* ability to place retirees in a commercial bank. In the virtual absence of government-supported social security, and in the fact of mandatory retirement at age 55, the retiree has to find suitable arrangement for another twenty-five to thirty years. Owing to the centrality of the commercial bank to the life of the principal and satellite firms of each trading group of partners à la *zaibatsu* of the earlier days, a career with a major bank is highly prized in Japan—and only MITI can place retirees in a commercial bank. Doubtless, this power of retirement placement was partially instrumental in MITI's success in persuading the recalcitrant shipbuilding industry that it should rationalize. In the preceding section, we noted that MITI got the Japanese Development Bank to grant a five-year moratorium on interest payments to the shipbuilding industry and that ninety-seven companies in that industry towed the line and allowed themselves to be consolidated into six groups. We could incorporate the power of placement of MITI into the CONCOL on rationalization:

- Country (MITI) alter ego *with* rationalization *with* bank moratorium on interest payments *with* retiree placement in commercial banks
- Country (MITI) anti-ego *against* rationalization *against* bank moratorium on interest payments *against* retiree placement in commercial banks.

0.0 Commercial Bank's Support of Long-term Planning

Even as MITI takes a long-term view of the prosperity of Japanese export markets when urging the leading auto manufacturers to set up plants in the United States, so also the bank that finances the companies takes a long-term view of the company operations. The local bank manager has considerable discretionary powers in lending money to the company. As Gibney points out, if a long-range plan is explained to the bank(s) that in their judgement makes sense, the banks will be willing to wait for a long time for the profits. As Chalmers Johnson put it, the Japanese economy is *plan-rational*, while the American economy is *market-rational*.[11] We can use these terms to identify the CONCOL players:

10. Haitani, *op. cit.*, 137.
11. Chalmers Johnson, *MITI and the Japanese Miracle: The Growth of Industrial Policy, 1925–1975*, (Stanford: Stanford University Press, 1982).

- Company (commercial bank) alter ego *with* long-term plan-rationality (growth) *against* short-term market-rationality (profits)
- Company (commercial bank) anti-ego *against* long-term plan-rationality (growth) *with* short-term market-rationality (profits).

0.0 Commercial Bank's Rescue of Failing Company

Because of the plan-rational view of the bank, it also seeks direct knowledge of the operations of the company. Representatives of the bank sit on the board of directors of the company, and, unlike the American counterpart, are active decision-makers, not a one-day-a-year expert, mostly rubber-stamping what the operating officers want done. If the long-term plan is sensible, the bank will step in to save a company that is failing by replacing the management:

- Company (commercial bank) alter ego *with* long-term survival of the company *against* unsuccessful (board) policies
- Company (commercial bank) anti-ego *against* long-term survival of the company *with* unsuccessful (board) policies.

0.0 Family Savings Generating Loanable Funds

The commercial bank is able to encourage the companies to take a plan-rational approach which will bear fruit in the long-run, because the Japanese save nearly four times as much as the Americans. Clearly, the opposing interests, personified as players, are the immediate consumption and the deferred consumption:

- Wage earners alter ego *with* deferred consumption *against* immediate consumption
- Wage earners anti-ego *against* deferred consumption *with* immediate consumption.

0.0 Increased Cash Flow for Company Expansion

While the commercial bank provides the company with capital, lending four to five times its equity, the company enhances its cash flow by paying lower dividends to shareholders—38 percent of after-tax earnings compared to 52 percent in the United States. Further, the government allows the companies to retain higher cash flow for expansion by deferring taxes on retirement funds, overseas investment losses, accelerated depreciation, and reserves. We can identify two different sets of CONCOL players:

- Company alter ego *with* long-term growth *against* higher short-term dividends
- Company anti-ego *against* long-term growth *with* higher short-term dividends
- Government alter ego *with* deferred taxes *against* short-term revenue increase

229

- Government anti-ego *against* deferred taxes *with* short-term revenue increase.

0.0 Summary of Investmental Sets of CONCOL Players

The fundamental characteristics of the investmental factors of Japanese productivity, like that of the institutional factors, is the primacy of the long-term.

MITI selects and promotes industries that it considers most important to the long-term growth and prosperity of Japan. By its selection of heavy and chemical industries in the 1950s and 1960s and knowledge industries in the 1970s and 1980s, MITI necessarily de-emphasized other industries. Since computers have been supported by MITI since the 1950s, the 1980s emphasis of VLSI could be interpreted as de-emphasizing mainframe computers and emphasizing microcomputers.

In addition to its role as the guiding hand in identifying and sponsoring growth industries, MITI also determines the needs of steel users. To the extent that MITI's documents project the user's needs, the producers are supposed to meet the user's needs, not vice versa. Looking at the global users of Japanese automobile products, MITI has been urging Japanese automakers to set up manufacturing plants in the United States, a proposition that met stiff resistance for some time before slowly being adopted. To keep the Japanese industry healthy, MITI finds itself urging the merger of different companies in the same industry, again another proposition that met stiff resistance, but was finally carried out in several industries. One of the elements of persuasion that MITI exercised in bringing about rationalization was the moratorium on interest payments that the Development Bank of Japan extended to the shipbuilding industry, which did not want to have its companies merge. Part of the success of MITI in inducing the bank to take the step of moratorium on interest payments is traceable to the exclusive ability of MITI to place retirees in commercial banks.

Japanese companies, unlike their American counterparts, are under no pressure to make themselves look good in the next quarterly report. The bank that holds about 5 percent of the voting stock of the company takes a hard look at the long-term plans of the company; and if the plans are credible and the bank likes them, the bank will wait a long time for the profits. Should the company be failing during the execution of the long-term plan, the bank steps in to rescue the company, if necessary by replacing the management. The CONCOL players represent the opposing views of the short-term and the long-term.

So do the Japanese wage earner's personified alter ego and anti-ego. The deferred consumption wins hands down, boosting the Japanese savings to 22 percent of disposable personal income. The Government also opts for deferred revenue by allowing companies to raise their cash flow for expansion through deferred taxing of reserves.

The CONCOL players in the investmental arena of Japanese industry and

business explicate the choice made by the country for *long-term growth in preference to short-term profits.*

0.0 CONCOL in the Inventional/Innovational Factors of Japanese Productivity

Among the inventional/innovational factors that account for Japanese productivity, the most significant is the market perspective, which is served by accenting the positive in competitive advantage, both in production and in products.

0.0 Acquiring the Market

Gain the whole world market first is the practical philosophy of Japanese business. Once the market is gained, it is held by productivity improvement. At the initial phase, however, if there had to be a choice between market and productivity, market wins hands down. Gibney compares Japanese exports activity to an army on the march:

- Multinationals—test markets; arrange local contacts; buy up raw materials
- JETRO (Japan External Trade Organization)—intelligence on sales
- Government—tax credits, depreciation allowances, R & D assistance.

The world of the 1980s is not a homogeneous market, but a diverse and fragmented one. It is not a single outlet for a single product. Even with a single product, the needs it satisfies are different, requiring careful attention to product differentiation by customer needs. It is this market segmentation that underlies the Japanese strategy of seeking first the market. Ohmae points out:

> We have already seen that there are two types of segmentation: the first by *customer wants*, the second by *market coverage*. Correspondingly, market segments may undergo two possible types of structural change: one owing to the changes in user objectives over time, the other owing to the changes (either geographic or demographic) in the distribution of the user mix. Structural change normally obliges the corporation to shift its resources either across functions or across product-market segments (e.g. from Northeast to South Central). . . .
>
> By influencing the user's objectives, fundamental forces like these [e.g., convenience, economy and utility instead of high-speed travel and prestige in auto sales] create a multitude of replacement opportunities . . . By offering differentiated products that will have a stronger appeal to the new breed of customers than to the old, it can take

> advantage of the forces at work in ways most likely to
> enable it to grow faster, and more profitably, than its
> competitors.[12]

In general, the Japanese strategy is to find that particular segment of the differentiated product that will offer a special niche that the corporation can fill. Quite often, Los Angeles is a test market, the differentiated product being introduced and market-tested to determine how viable the market would be.

The CONCOL players reflect the bidding by the new (novel) product to supplant the old:

- Company alter ego *with* new customer wants *with* differentiated market segment

- Company anti-ego *against* new customer wants *against* differentiated market segment.

0.0 Accenting the Positive in Functional Competence

Ohmae points out that the secret of Japanese success lies in the *sequencing* of the improvement of functional competence. The idea is to select one particular segment among the many, and by concentrating all the possible resources in establishing pre-eminence in that particular segment, achieve competitive advantage over the competition. Manufacturing engineering was the area of functional competence chosen by many Japanese companies in the 1950s and 1960s. Subsequently, other areas of emphasis were chosen: first, quality control and product design; later, basic research and direct marketing. It is the *sequence* that is the secret. The CONCOL players would naturally opt for opposing sequences at a given time:

- Company alter ego *with* manufacturing engineering emphasis *against* quality control emphasis

- Company anti-ego *against* manufacturing engineering *with* quality control emphasis.

At a later time:

- Company alter ego *with* basic research and direct marketing emphasis *against* quality control emphasis

- Company anti-ego *against* basic research and direct marketing emphasis *with* quality control emphasis.

0.0 Accenting the Positive in Customer Want Satisfaction

Turning from functional competence in services, such as manufacturing engineering, to the products themselves, the most sensitive issue is that of corporate identity. What Canon, Rioch, Panasonic, and Pentax found was that they needed to sell

12. Ohmae, *op. cit.* 104–6.

not under their own but under foreign tradenames. By selling under foreign trade-names, they could ensure that their products indeed satisfied specific customer wants. Acquiring enough market validation through other brand names, they could then sell under their own brand name. The CONCOL players would opt for disclosing corporate identity either early on or later:

- Company alter ego *with* foreign brand names initially *against* own brand names
- Company anti-ego *against* foreign brand names initially *with* own brand names.

0.0 Accenting the Positive in Cheaper Prices
Closely related to establishing the ability to meet customer wants is the objective of providing for these wants at cheaper prices. Matsushita aggressively pursues the strategy of lowering prices very quickly. Given that the customer is satisfied with the product, he would certainly buy from the one who sells it for less than for more. Matsushita, who rarely originates a new product, manages most of the time to make competitive products for less than the competition. Cheaper prices compete against higher markups. Matsushita's philosophy is that smaller markup on a large and expanding market is better than larger markup on a small and shrinking market. So the CONCOL players here opt either for larger or smaller markups:

- Company alter ego *with* smaller markups *with* smaller prices *with* larger market
- Company anti-ego *against* smaller markups *against* smaller prices *against* larger market.

0.0 Accenting the Positive in Faster Product Life Cycle
While Matsushita produces the same product (more or less) as the competition and lowers the prices, Casio thrives on making a *different product* much faster than the competition, then lowering the price. First, reduce the price of the same (or similar) product as the competition (e.g., 2 mm-thick card-size calculator); second, quickly introduce a different product (e.g., a new model which plays musical notes) and watch the competition take another two years to catch up.

There is a clear advantage in shorter product life cycles that turn out new products faster. However, the customer's hunger for the novel, may not continue unsatisfied forever. Moreover, they may resent the fact that no sooner have they bought something than it is obsolete. A judicial balance between accelerating the novelty of products and continuing serviceability, including maintenance service, is required. The CONCOL players here would represent opposing views on the length of the product life cycle:

- Company alter ego *with* shorter product life cycle *with* faster obsolescence *against* longer product life

233

- Company anti-ego *against* shorter product life cycle *against* faster obsolescence *with* longer product life.

0.0 Innovation for "Self-Renewal"

Which level of the company should be responsible for assuring not a single product, but a stream of new products, to replace the older ones?

Insofar as the operating divisions are closest to the customer, they would know the customer preferences. They could, therefore, suggest modifications of current products and/or services which would retain and expand the market.

To meet the customer wants, however, particularly in the future, it is necessary to invest in R & D, maching tooling and design, product design, etc. If the division is measured on contribution to profits in the short-term, then the division will not want to sacrifice present earnings for potential earnings.

One way to reconcile the conflict between the short-term profits and the long-term growth is to institute two reporting relationships, one measuring primarily the short-term, the other measuring primarily the long-term. This is done at Matsushita by having the division manager reporting horizontally to the vice president for product groups and vertically to the president of the company.

> In 1953 he [Matsushita] organized his divisions into product groups with division managers reporting vertically to the President of Matsushita and horizontally to their group Vice-Presidents, who served as specialists with detailed knowledge on a whole family of products. This concept of having two bosses was anathema to managers of his day. Matsushita reckoned, however, that we all grow up under two bosses—a mother and a father—and that it is the nature of life to have to juggle the complexities that arise from such arrangements. . . .
>
> Then came the early sixties and another period of recession and stagnation. This time, Matsushita *decentralized* to give more initiative to the field; for the first time, each group product was given full responsibility for its marketing and sales activities.[13] (emphasis added)

The division retains 40 percent of its pretax profits for its self-renewal, i.e., bringing on line a new generation of products to replace the older ones. In addition, the corporate headquarters can help launch major initiatives from the fund created out of the 60 percent of division profits. The CONCOL players point up the conflict between new product initiatives and short-term profits:

13. Richard T. Pascale and Anthony G. Athos, *The Art of Japanese Management*, (New York: Warner Books, 1982), 45–47.

- Company alter ego *with* product initiatives *against* short-term profits
- Company anti-ego *against* product initiatives *with* short-term profits.

0.0 Corporate Controls of Division Performance

It is one thing to *plan* for self-renewal; it is quite another to implement it.

To ensure that the divisions are kept on the straight and narrow path of short-term profits without sacrificing long-term growth, so that they sell the present products well even as they create and introduce new products, three sets of plans —five-year, two-year, and six-month—are required to be presented and monitored. Of the three, the six-month is the one most rigorously monitored. The monitoring is done by corporate controllers who exercise a very special relationship of conflict without conflicting. It is interesting to note that the Matsushita controller uses an unusual term to refer to his role:

> We are rarely viewed as watchdogs—partly because Matsushita's comprehensive system encourages a very open operation. . . . We act more like a wife in the traditional Japanese household. Like the wife, we work largely outside the public view but keep tabs on the finances and remind the division head how things are doing.[14]

"How things are doing" covers the gamut of differences between the detailed six-month operating plan and the actual performance. What the controller does is to understand how the variances occur, anticipate the more troublesome variances, and discuss with the division manager how they may be averted or corrected. The role of the controller could be viewed as being in conflict with that of operations, although both are on the same team:

- Company alter ego *with* minimum variances *with* short-term performance goals
- Company anti-ego *against* minimum variances *against* short-term performance goals.

0.0 Summary of Inventional/Innovational Sets of CONCOL Players

The fundamental characteristic of the inventional/innovational factors of Japanese productivity, like that of institutional and investment factors, is the primacy of the long-term growth.

To survive and grow in the long-term, the Japanese concentrate first on the global market. They seek to influence the user's objectives and segment carefully

14. Naoyumi Tsumoga, "Accounting Functions of Matsushita," in *Business Behavior, op. cit.*, 205.

Robotics/Artificial Intelligence/Productivity

the market for each element of the product. With a clear-cut idea as to who the customer is and what he wants, the company decides on the *sequencing of elements of functional competence* that would yield the most competitive advantage. In general, the sequencing has been: manufacturing engineering, quality control, basic research, and direct marketing. While these functions give the company an edge over competition, especially in the *services* it can offer the customer, the *products* themselves are offered in foreign markets under foreign brand names in the initial stages of marketing by some of the well-known manufacturers like Canon, Panasonic, etc. Once the customer preference for a particular product is established, the prices are lowered as soon as possible in a strategy to deny entry by competition. While the price reduction applies to the same product, *different* products could also be offered in quick succession (e.g., Casio). In introducing new products there is the conflict between the long-term growth of the company and the short-term profits; which requires the division manager to report to different levels of higher management, focusing on the short-term and the long-term. The plans for the next six months, for both present and potential products, are carefully scrutinized by the corporate controllers at Matsushita, so that they can alert the division managers to potential deviations from the plans, which may be averted or corrected.

The CONCOL players personify the opposing interests affecting Japanese productivity. The opposing interests are harmonized through the CONCOL process, which we will consider in chapter 11.

10

CONCOL Players in Robotics in the United States

OVERVIEW: While in Japan, the alter ego in the institutional, investmental, and inventional/innovational elements is more often than not successful in having its long-term planning view prevail over the anti-ego's short-term profits view, it is not so in the United States.

Overtaken by the Japanese in quality product at cheaper prices, the U.S. has already lost sizeable segments of the domestic market to Japan: more than a fifth of the automobile market, more than a third of the camera market, more than half of the watches and radios market, just to mention a few. Yet the opposing concerns of LABOR accent the clinging to present jobs much more than the active preparation for new jobs, even though in less than two decades, manufacturing could be employing 2 percent instead of 27 percent of the workforce. Instead of producing goods through manufacture, information is produced for internal and external consumption in the information economy, which in 1967 produced 53 percent of the income earned. While the number of information workers increased by 83 percent between 1960 and 1980, the blue collar workers and farmworkers increased by less than half as much at 36.4 percent. Nearly 90 percent of the new jobs during the 1970s were *not* created in the goods-producing sector.

Turning from labor to CAPITAL, the opposing concerns again accent the short-term profits, most often displacing long-term growth as the primary objective. We discuss three elements of profitability and four elements of productivity with reference to ten manufacturing industries, on the basis of which two industries emerge as clear candidates for robotization: motor vehicles and equipment, and primary metal industries. At the other end of the robotics candidate spectrum are two industries which can well afford robotics: instruments and related products, and machinery, excluding electrical.

Turning from capital concerns of robotization to TECHNOLOGY concerns, we find that, as late as October 1982, American industry entertained notions of piece-meal robotics, with an NC (Numerical Control) here and a DNC (Direct Numerical Control) there. In direct contrast is the wholesale robotics approach taken

by Japan's Ohgishima steel mill which in 1980 produced eighteen hundred tons per worker, *more than three times* that of Bethlehem Steel Company's modern Burns Harbor plant. The wholesale robotics, which use only a third of the labor employed by comparable steel mills, was introduced with the advance concurrence of labor, which participated in the detailed training/transfer of affected workers. From the initial discussion of CONCOL players *within* Japan and *within* the United States, we turn to CONCOL *between* the United States and Japan, seeking a possible basis for simultaneous competition and cooperation in identifying and acquiring selective market segments at home and abroad, using high-technology investments, inventions and innovations.

0.0 Secure the Short-Term; Sacrifice the Long-Term

To robotize or to retrogress is no longer a rhetorical question; it is an imperative issue.

Overtaken by the Japanese in quality products at cheaper prices, the U.S. has already lost sizeable segments of the *domestic* market to Japan: more than a fifth of the automobile market, more than a third of the camera market, more than half of the watches and radios market, to mention a few. To survive, U.S. industry has not only to meet, but also exceed, the Japanese performance in quality and price. Japan's gains in both quality and price are due in large measure to robotization.

More important than the actual extent of robotization is the long-term commitment of Japan to do whatever is necessary to make Japan the leader in chosen field(s) of manufacture and exports. Industry is fully supported by government in achieving the national goals of economic superiority world-wide and extending markets world-wide.

By comparison, U.S. industry is characterized by the myopia of the next quarterly profit. Since Wall Street, and not banks, provide capital to U.S. industry, quarterly performances reflected in the company's stock prices exercise almost life-and-death powers. Reginald Jones, former chairman of the board and chief executive officer of General Electric, said:

> Too many managers feel under pressure to concentrate on
> the short-term in order to satisfy the financial community
> and the owners of the enterprise—the stockholders. In
> the United States, if your firm has a bad quarter, it's
> headlines. Real trouble ensues. The stock price falls out
> of bed. That's far different from Japan and Germany.[1]

1. Reginald Jones, cited in *Megatrends: Ten New Directions Transforming Our Lives*, by John Naisbitt (New York, New York: Warner Books, 1982), 81.

Table 10.1

The Factory Jobs That Will Be Directly Affected by Robots

Assemblers	1,289,000 employees
Checkers, Examiners, Inspectors, Testers	746,000
Production Painters	185,000
Welders and Flame-cutters	713,000
Packagers	626,000
Machine Operatives	2,385,000
Other Skilled Workers	1,043,000
	6,987,000

Source: "Changing 45 Million Jobs", *Business Week*, August 3, 1981, 67

The company is graded not only *externally* on its short-term performance, but also *internally*. The compensation plans for executives are, for the most part, tied to the performance in the short run. They would *lower* their income if they took a long-term view. So why should they look to the long-term?

To clarify the conflict between the short- and long-term interests of the same company, we may personify these interests, the alter ego representing the long-term interests of survival and the anti-ego representing the short-term profits. As stated earlier, we know that neither can long endure without the other. But we personify them to marshall the arguments on behalf of the two adversaries to consider the concomitant coalition players in robotics in the U.S.

We may group the major issues into three: labor concerns, capital concerns, and technology concerns.

0.0 Opposing Labor Concerns—Loss of Present Jobs vs. Promise of New Jobs

Just as in the early days of the Industrial Revolution, so also in the early days of the Robotics Revolution, a paramount concern is the imminent loss of jobs.

0.0 Robotics Displacement of Industry Jobs

A study by Carnegie-Mellon University concluded in 1981 that robots could displace workers in seven million existing factory jobs in the automotive, electrical equipment, machinery, and fabricated-metals industries. (See table 10.1.)

We see from table 10.2 that the number employed in Manufacturing Industries in 1981 was 20,173,000. The percentage of jobs affected directly by robotics works out to 34.64 percent, *more than a third*.

The replacement of 7 million workers as listed in table 10.1 is estimated to take place in twenty years by non-sensor-based robots, sensor-based robots, non-servo-controlled robots and servo-controlled robots.

Robotics/Artificial Intelligence/Productivity

At the October 1982 seminar, "Robotics and the Factory of the Future," Donald L. Smith, director of the Industrial Development Division of the Institute of Science and Technology at the University of Michigan, reported on an on-going Delphi study that the University is conducting jointly with the Society of Manufacturing Engineers:

> Current robot applications, in order of frequency, are esti-mated to be:
> - pick and place
> - machine loading
> - continuous path
> - manufacturing processes
> - inspection, and
> - assembly.
>
> By 1985, machine loading and assembly applications will become more prevalent. In 1990, *assembly* robots are ex-pected to lead in the *electronics* industry and be secomd in *automotive* and *light manufacturing*. Machine loading will lead in *heavy manufacturing* and in *casting and foundry* applications.[2] (emphasis added)

Assembly offers large cost savings and is therefore likely to receive increased attention in the immediate future, enabling machine loading and assembly applications to become prevalent by 1985, as the Michigan Delphi predicts. An earlier forecast from the Delphi predicts that by 1987, 15 percent of all manufacturing assembly systems will use robotics.[3]

We see from table 10.2 that in thirty years, the percentage of those employed in manufacturing declined from 38 percent to 27 percent in 1981. The Carnegie-Mellon and Michigan University predictions of drastic reductions in manufacturing employment reminds us of the impact of the Industrial Revolution on American agriculture. As late as the turn of the century, more than a third of the labor force was employed in agriculture, which now employs about 3 percent of the labor force. By the turn of the twenty-first century, manufacturing could be employing less than 2 percent of the labor force, according to an executive at Cincinnati Milacron, a major U.S. robot manufacturer.[4]

---

2. Rita R. Schrieber, "Robotics in the Eighties," *Robotics Today*, October 1982, 41.
3. John Teresko, "Stirring Interest in Robot Technology," *Industry Week*, Vol. 203, No. 124, 26 November 1979.
4. Michael Funke, "The Invasion of Job Eating Robots," *Labor Today*, Vol. 19, No. 4, December 1980.

0.0 Creation of New Jobs by the Information Economy

What is the impact of the reduction of the percentage of the labor force employed in agriculture? What parallel inference may be drawn about a similar possible drop in the percentage employed in manufacturing?

The reduction in the labor force engaged in agriculture did not reduce the output of agriculture. On the contrary, it was precisely because the output could be increased multifold by using implements of the Industrial Revolution that fewer and fewer farmers could turn out greater and greater agricultural products.

Similarly, when the percentage of labor employed in manufacture declines from over 20 percent to under 2 percent, it would be because the output of manufactured goods and services could be increased multifold using implements of the Robotics Revolution, enabling fewer and fewer manufacturing workers to turn out greater and greater products of manufacture.

When the farmer no longer had to work from sunup till sundown to eke out a living for himself and his family, but could produce enough to sell to others, the younger generation migrated to the cities, looking for work with the new implements in the factory. Shorter working hours and better working conditions (sans mother's good cooking) opened up new vistas of life, with discretionary hours in which one could enlighten, entertain, and educate oneself. Book learning could help move one up the ladder of factory employment, including managerial positions.

This change in occupational structure over the years is reflected in the respective private sector shares of "goods producing" and "service-producing" industries. In 1950, 47 percent were engaged in goods production; in 1981, only 33 percent. In the meantime, 53 percent were engaged in service industries; in 1981, 67 percent.

In table 10.3, we present the GNP by industry—*private sector*. Using both tables 10.2 and 10.3, we find that in 1960, 46 percent of the labor force produced 42 percent of the GNP in goods; in 1970, 39 percent of the labor force produced 40 percent of the GNP in goods; in 1979, 33 percent of labor produced 30 percent of the GNP.

This shift in industrial mix, as we have seen in chapter 2, is a major source of productivity decline in the United States, accounting for as much as 22 percent of the decline in productivity growth in 1972–78, compared with 1948–65.

Now we are poised at the start of another shift in industrial mix to generate what Marc Porat called in 1977 the information economy. Porat studied 440 occupations in 201 industries to identify the information jobs and to compile their contribution to GNP. For 1967, he calculated as follows:

> Primary Information Sector   25.1% of GNP
> (Clerks, Librarians, Systems Analysts, etc. employed in
> information industries, such as computer manufacturing,
> telecommunications, printing, mass media, advertising,

Table 10.2
Private Sector (Payroll) by Industry
(Employment in Thousands)

Industry	1950		1955		1960		1970		1981	
Goods-Producing		47%		47%		46%		39%		33%
Mining	901	2%	792	2%	712	2%	623	1%	1,132	1%
Construction	2,364	7%	2,839	7%	2,926	6%	3,588	7%	4,176	5%
Manufacturing	15,241	38%	16,882	38%	16,796	38%	19,367	31%	20,173	27%
Service-Producing		53%		53%		54%		61%		67%
Transportation	4,034	10%	4,141	9%	4,004	9%	4,515	8%	5,157	7%
Wholesale trade	2,635	7%	2,926	7%	3,143	7%	3,993	7%	5,359	8%
Retail trade	6,751	18%	7,610	17%	8,248	17%	11,047	20%	15,192	20%
Finance, insurance	1,888	4%	2,298	6%	2,629	6%	3,645	6%	5,301	7%
Services	5,357	14%	6,240	14%	7,378	15%	11,548	20%	18,592	25%
Total	39,170	100%	43,727	100%	45,836	100%	58,325	100%	75,081	100%

Basic Source: Monthly Labor Review, January 1983, 77.

accounting and education; risk-management industries, including finance and insurance)

Secondary Information Sector   21.1% of GNP
(Information workers employed in non-information industries, such as goods-producing industries)

Porat calculated that the information economy accounted for some 46 percent of GNP and more than 53 percent of income earned.[5]

The core of Porat's classification is information: people producing information for internal consumption within as well as without, whether they are employed in the primary or in the secondary information sector. The increasing accent on information is reflected, in table 10.4, in the rise in the number of people working in information generation.

The highest percentage of employed workers in clerical workers (18.6%), followed by professional and technical workers (16.1%), followed by managers and administrators (11.2%), accounting together for a total of 45.9% of the work force in 1980. This total share represents a 24.3% increase over its corresponding value in 1960 of 36.9%. The rise is even more impressive in terms of the actual number of workers in the three categories. In 1960, the total number was 24,298, which increased 83.7% to 44,637 by 1980. The information economy is, indeed, clearly perceptible as a reality, and not a mere abstraction.

While the number of information workers increased by 83.7% between 1960 and 1980, the blue collar workers and farmworkers increased, at 36.4%, by less than half as much.

David Birch's paper, "Who Creates Jobs?" shows that of the nineteen million jobs created during the 1970s, only 5 percent were in manufacturing and only 11 percent in the goods-producing sector as a whole. Nearly 90 percent of the new jobs were *not* created in the goods-producing sector.[6]

0.0 Labor CONCOL in Robotics

To highlight the inherent opposition in the interests of labor with respect to robotics, we can personify the opposition in terms of alter ego and anti-ego:

- Labor alter ego *with* robotics *with* new information economy jobs *against* present jobs

- Labor anti-ego *against* robotics *against* new information economy jobs *with* present jobs.

Following the onset of the Industrial Revolution, it took ten years before more jobs were created by the expanding demand for British exports than were lost to spinning jennies. There were among those who lost their jobs some who never

5. Marc Porat, Information Economy: Definition and Measurement, (Washington, D.C.: U.S. Department of Commerce, Office of Telecommunications, May 1977).
6. David Birch, "Who Creates Jobs?", *The Public Interest*, Fall 1981.

Table 10.3
Private Sector Gross National Product by Industry

Industry	1960		1970		1979	
Good-Producing		42%		40%		30%
Mining	14.4	3%	19.1	2%	21.2	2%
Construction	46.1	8%	57.1	7%	59.3	5%
Manufacturing	172.0	31%	260.6	31%	352.0	31%
Service-Producing						
Transportation	30.9	5%	43.8	5%	55.9	5%
Wholesale trade						
Retail trade	117.9	20%	178.4	22%	243.3	22%
Finance, insurance	101.9	18%	152.9	18%	222.6	19%
Services	82.2	15%	124.7	15%	174.4	16%
	565.4	100%	836.6	100%	1,128.7	100%
	= 77% of		= 78% of		= 80% of	
All industries	736.8		1,075.3		1,431.6	

Basic Source: U.S. Bureau of Economic Analysis, *The National Income and Product Accounts of the United States, 1929–74*, and *Survey of Current Business*, July issues, April 1980.

found suitable employment again. The labor anti-ego position is really the position of opting for survival in the present, holding on to present jobs, rather than hoping for the future jobs. Unless labor anti-ego resorts to Luddite-type destruction of robots, some will probably be unemployed and be unemployable with the change-over to robotics.

0.0 Conflicting Capital Concerns—Short-Term Profits vs. Long-Term Growth

The CEO, with an average tenure of five years, has little motivation to undertake long-term investments that will probably not bring results for ten years or more. He will therefore necessarily take as short-term a view as is consistent with the incentive structure that rewards him. Since the next quarterly report can make him quite vulnerable, he will naturally seek to achieve results in that period.

If a plant or operation is unprofitable, the effort required to make it profitable may well go beyond the next quarter(s). A major goal is the increase of productivity by investing in machinery at a significantly higher initial cost.

Such a long-term investment may well be the way to increase productivity, reduce costs, and increase the market for the outputs of the company, ensuring its growth in the long-run. This objective may be associated with the alter ego of the company, and the opposite objective associated with the anti-ego of the company.

Table 10.4
Occupation of Employed Workers: 1960 to 1980

	1960	1970	1980	1960	1970	1980
Total ('000)	65,778	78,627	97,270	100.0%	100.0%	100.0%
White-Collar workers	28,522	37,997	50,809	43.4%	48.3%	52.2%
Clerical workers	9,762	13,714	18,105	14.8%	17.4%	18.6%
Professional and technical	7,469)24,298	11,140)33,143	15,613)44,637	11.4%)36.9%	14.2%)42.1%	16.1%)45.9%
Managers and administrators	7,067	8,289	10,919	10.7%	1.05%)	11.2%
Salesworkers	4,224	4,854	6,172	6.4%	6.2%	6.3%
Blue-Collar workers	24,057	27,791	30,800	36.6%	35.3%	31.7%
Craft and kindred workers	8,554	10,157	12,529	13.0%	12.9%	12.9%
Operatives excl. transport	11,950	13,909	10,346	18.2%	17.7%	10.6%
Transport equipment oper.			3,468			3.6%
Nonfarm laborers	3,553	3,724	4,456	5.4%	4.7%	4.6%
Service workers	8,023	9,712	12,958	12.2%	12.4%	13.3%
Farmworkers	5,176	3,126	2,704	7.9%	4.0%	2.8%

Basic Source: U.S. Bureau of Labor Statistics, *Employment and Earnings, Monthly.*

- Company alter ego *with* long-term growth *against* short-term profits *with* capital investment
- Company anti-ego *against* long-term growth *with* short-term profits *against* capital investment.

0.0 Closing the Ailing Plant

The industries in which productivity-improving machinery investment is needed are those in which the world-wide market for the company products has been significantly impaired. With Japan capturing over a fifth of the automotive market, automotives is an industry that qualifies under this criterion. Similarly, electrical and electronic appliances qualify.

The resources to be committed are significant in size and in duration. Further, the reaction from labor unions in the industry is a serious consideration. Should the company find that *a*, the capital costs are too high, *b*, the labor costs in terms of compensation for loss of jobs or for transfers, etc. are too high, or *c*, labor's reaction to robotics is anywhere from reluctant to resistant, then it may find that closing down a plant is a more attractive solution than fighting for long-term growth and prosperity of the company. By closing down an ailing plant, the write-offs would probably improve the after-tax position of the company on the next quarterly report.

- Company alter ego *with* capital investment *against* plant closing *against* short-term profits *with* long-term growth
- Company anti-ego *against* capital investment *with* plant closing *with* short-term profits *against* long-term growth.

0.0 The Closing of GM's Fremont Automotive Plant

In 1982, General Motors closed down their Fremont, California automotive plant, which employed three hundred people. In 1983, it was announced that the plant would reopen under GM-Toyota ownership. The laid-off employees assumed that they would be the first to be recalled, but the Japanese management of the plant let it be known that (1) the new plant would entail heavy investment in robotics, (2) requiring special skills on the part of the employees, (3) who would be significantly fewer than the three hundred, and (4) who would best be chosen on the basis of new skills and outlook than old skills and habits appropriate to an earlier day.

The Fremont plant *reopening* under Toyota-GM management is a CONCOL *between* the United States and Japan, while the plant *closing* would be a CONCOL *within* the United States.

0.0.0 CONCOL Within the United States on Plant Closing

- Company alter ego *with* continued plant operations *against* plant closing *with* long-term growth *against* short-term profits.

- Company anti-ego *against* continued plant operations *with* plant closing *against* long-term growth *with* short-term profits
- Labor alter ego *with* continued plant operations *against* plant closing *with* long-term growth *against* short-term gains
- Labor anti-ego *against* continued plant operations *with* plant closing *against* long-term growth *with* short-term gains.

We see that the alter egos of both company and labor are for continued plant operations, the only difference being their respective opposition to short-term profits and short-term gains. In other words, the long-term interests of both the company and labor are to forego short-term advantage in the form of higher wages/benefits for the latter, and tax write-offs/higher quarterly profits for the former.

0.0.0 CONCOL Between Japan and the United States on Plant Reopening

Why did Japan enter into the agreement with GM?

There was obviously a *quid pro quo* from both points of view. Japan found that in the land-sparse homeland, it would cost $1.5 billion to duplicate the plant that GM could offer. For its part, GM found that Japan would invest $1.5 billion in robotics to make possible the production of quality cars at cheaper prices. Further, GM would like to observe first-hand how Japan manages people-assets.

The best intentions of Toyota and GM would come to naught unless there were an assured market for what the new plant—rather, the old plant under new management—would produce. Knowing that the world-wide market for automobiles is not unlimited, Japan felt that it should cooperate with GM to share the market with them, rather than compete with each other and lose the market altogether.

- Japan (Toyota) alter ego *with* long-term growth of market *against* short-term competitive profits *with* robotics investment *with* Fremont plant reopening
- Japan (Toyota) anti-ego *against* long-term growth of market *with* short-term competitive profits *against* robotics investment *against* Fremont plant reopening
- U.S. (GM) alter ego *with* long-term growth of market *against* short-term competitive profits *with* robotics investment *with* Fremont plant reopening
- U.S. (GM) anti-ego *against* long-term growth of market *with* short-term competitive profits *against* robotics investment *against* Fremont plant reopening.

0.0 Affected Industry Market—Lost Domestic Market

From the point of view of the United States alter ego, what industries would it need to inject capital resources for robotics?

Robotics/Artificial Intelligence/Productivity

Ideally, it would be in those industries whose world-wide markets are weakest, assuming that every segment is equally desirable. One index of the weakness of the world market would be the shrinking *domestic* market. Considering only the *Japanese* inroads into the U.S. domestic market, the industries requiring injection of capital resources for robotics are, in descending order of Japanese share of U.S. consumption:

1. Motorcycles   90%
2. Watches and Radios   55%
3. Cameras   35%
4. Automobiles   25%
5. Steel   15%

While the loss of the domestic market to Japan in the case of motorcycles is indeed high (or, from the U.S. alter ego point of view, the U.S. share of the motorcycle market is -0.90), that fact alone would not cause the U.S. alter ego to concentrate all U.S. capital resources on recapturing the lost domestic market from Japan, including investing heavily in robotics.

0.0 Affected Industry Assets—Share of Total Assets

Using U.S. Internal Revenue Service figures,[7] we find that the total assets of active corporations in 1977 was $5,326.4 billion. Of this total, manufacturing industry owns $1,182.3, or 22.20%. All the five products with significant *negative* U.S. markets are included in the category of manufacturing industry. In descending order of industry share of total assets are:

1. Finance, insurance, real estate   53.73%
2. Manufacturing   22.21%
3. Transportation and public utilities   10.12%
4. Wholesale and retail trade   7.79%
5. Services   2.24%
6. Construction   1.71%
7. Mining   1.66%
8. Agriculture, forestry, and fishing   0.54%
                                          100.00%

0.0 Affected Industry Profitability—Corporate Profits

The commitment of capital to robotics would be easier in industries that are highly profitable. Based on the *Survey of Current Business* data[8], the total corporate

7. U.S. Internal Revenue Service, *Statistics of Income, Corporation Income Tax Returns, 1977*, (Washington, D.C.: Government Printing Office, 1978).
8. *Survey of Current Business*, July 1981.

profit for all industries in 1977 was $167.8 billion, with manufacturing industries in the lead with 37.43% in descending order:

1.	Manufacturing	37.43%
2.	Wholesale, and retail trade	15.85%
3.	Finance, insurance	9.78%
4.	Communication	8.46%
5.	Services	3.99%
6.	Transportation	2.20%
7.	Construction	2.03%
8.	Mining	1.67%
9.	Agriculture	.36%
	Rest of the world	18.23%
		100.00%

0.0 Affected Industry Profitability—Profits per Dollar of Sales

More than the total value of profits, it is the profitability of the particular business that could influence capital resource commitment. The *Quarterly Financial Report for Manufacturing, Mining and Trade Corporations* shows that all manufacturing corporations together earned 4.9¢ per each dollar of sales in 1980.[9] The profits per dollar of sales of selected industries are as follows:

ABOVE 4.9%			
	1.	Instruments and related products	9.3% per dollar sale
	2.	Machinery, excluding electrical	6.5%
	3.	Nonferrous metals	5.9%
	4.	Electrical and electronic equipment	5.0%
	5.	Aircraft, guided missiles and parts	4.3%
BELOW 4.9%	6.	Fabricated metal products	4.2%
	7.	Stone, clay, and glass products	4.2%
	8.	Primary metal industries	4.1%
	9.	Iron and steel	2.9%
	10.	Motor vehicles and equipment	− 3.4%

0.0 Affected Industry Profitability—Profits as % of Stockholders' Equity

How much return does the stockholder get in 1980? All manufacturing corporations earned 13.9% of stockholders' equity as profit.[10]

9. U.S. Federal Trade Commission, *Quarterly Financial Report for Manufacturing, Mining and Trade Corporations.*
   10. *Ibid.*

Robotics/Artificial Intelligence/Productivity

AT OR ABOVE 13.9%	1. Instruments and related products	17.5%
	2. Aircraft, guided missiles and parts	16.0%
	3. Nonferrous metals	15.6%
	4. Electrical and electronic equipment	15.1%
	5. Machinery, excluding electrical	15.0%
	6. Fabricated metal products	13.9%
BELOW 13.9%	7. Primary metal industries	11.7%
	8. Stone, clay, and glass products	10.8%
	9. Iron and steel	9.0%
	10. Motor vehicles and equipment	-9.3%

0.0 Affected Industry Productivity—Capital Stock per Hour Worked

Profitability measures reflect the use of capital stock. Since robotics will increase the capital stock per hour worked, the present status can be looked upon as a starting point. Public utilities had the highest ratio of capital to labor in 1979 at 167.09 and the apparel industry, the lowest, at 2.98.[11]

ABOVE 10.00	SIC Code	49	1. Public utilities	167.09
		29	2. Petroleum	67.60
		48	3. Communication	67.06
		41-47	4. Transportation	37.60
		21	5. Tobacco	35.93
		33	6. Primary metals	29.65
		28	7. Chemicals	28.73
		26	8. Paper	22.03
		10-14	9. Mining	21.93
		32	10. Stone, clay, glass	15.29
		37	11. Transportation equipment	13.48
		20	12. Food	13.06
		35	13. Non-electric machinery	11.72
		60-64	14. Finance and insurance	10.56
BELOW 10.00		22	15. Textiles	9.43
		34	16. Fabricated metals	9.30
		70-80	17. Services	9.21
		30	18. Rubber and plastics	8.96
		36	19. Electric machinery	8.47

11. The Conference Board, *Economic Road Maps*, New York.

24	20. Lumber and wood	7.64
38	21. Instruments	7.52
50-59	22. Wholesale, retail trade	7.33
27	23. Printing and publishing	7.01
25	24. Furniture and fixtures	5.02
31	25. Leather	3.58
15-17	26. Construction	3.39
3	27. Apparel	2.98

0.0 Affected Industry Productivity—Capacity Utilization

An index which combines the potential to manufacture (represented by the capital stock per hour worked) and the market is the capacity utilization rates. The *Survey of Current Business* provides quarterly data, indicating that all manufacturers operated at 78 percent capacity utilization in 1980.[12] The industries, in descending order of capacity utilization, are:

ABOVE 78	1. Machinery, except electrical	90
	2. Paper	88
	3. Petroleum	81
	4. Textiles	80
	5. Electrical machinery	80
	6. Aircraft	79
————————	7. Chemicals	78
BELOW 78	8. Food, including beverages	76
	9. Primary metals	74
	10. Stone, clay, and glass	73
	11. Rubber	70
	12. Motor vehicles	67

0.0 Affected Industry Productivity—Industrial Production

The productivity of the industry is going to be directly affected by the introduction of robotics. The *Federal Reserve Bulletin* shows the index of industrial production in 1980 at 147, to the 1967 base of 100.[13] In descending order of productivity are the industries as follows:

ABOVE 147	1. Rubber and plastics	256
	2. Chemicals and Products	207
	3. Electrical machinery	173

12. *Survey of Current Business*, monthly.
13. *Federal Reserve Bulletin*, monthly.

	4. Instruments	171
	5. Nonelectrical machinery	163
	6. Paper and products	151
	7. Furniture and fixtures	150
	8. Foods	149
	9. Clay, glass, stone products	147
	10. Printing and publishing	140
BELOW 147	11. Textile mill products	137
	12. Petroleum products	135
	13. Fabricated metal products	135
	14. Apparel products	129
	15. Tobacco products	120
	16. Lumber and products	119
	17. Transportation equipment	117
	18. Primary metals	102
	19. Leather and products	70

0.0 Affected Industry Productivity—Aging Equipment

The productivity of U.S. industry vs. Japanese industry (among others) depends on the relative ages of the equipment used by the respective countries. The Committee on Computer-Aided Manufacturing of the National Academy of Sciences says:

> Several studies have indicated that production facilities for batch manufacturing in the United States are significantly less modern than those in Japan and Germany. In the U.S., 28 percent of the metalworking equipment is more than 20 years old, compared with 23 percent in Japan and 7 percent in Germany. Worse, *only 33 percent* of the U.S. equipment is *less than 10 years old*, compared with *63 percent* in both *Japan and Germany*.[14] (emphasis added)

0.0 Capital CONCOL in Robotics

To highlight the inherent opposition in the interests of capital with respect to robotics, we can personify the opposition in terms of alter ego and anti-ego:

- Capital alter ego *with* long-term growth *against* short-term profits *with* capital investment

14. Committee on Computer-Aided Manufacturing, "Report on Activities January 1980 to June 1981," (Washington, D.C.: National Academy Press, 1981), 5.

▪ Capital anti-ego *against* long-term growth *with* short-term profits *against* capital investment.

Even as an abstraction, "capital" is too large a collective term. "Industry" or "company" would be more appropriate an entity of capital.

What "industry" would be most concerned about robotics?

We considered several relevant factors: (1) the lost domestic market (e.g., the automobile industry has already lost 25% of the domestic market to Japan), (2) the share of total corporate assets (e.g., manufacturing owns 22.21% of the corporate assets in the U.S.), (3) profitability—corporate profits (e.g., manufacturing earns 37.43% of total corporate profits), (4) profitability—profits per dollar of sales (e.g., instruments and related products earn 9.3¢ on dollar of sale), (5) profitability—profits as a percentage of stockholders' equity (e.g., instruments and related products earned 17.5% of stockholders' equity as profits), (6) productivity—capital stock per hour worked (e.g., instruments had a capital to labor ratio of 7.52), (7) productivity—capacity utilization (e.g., electric machinery operated at 80 percent capacity), (8) productivity—industrial production (e.g. instruments registered an index of 171 of industrial production in 1980 to a base of 100 in 1967), and (9) productivity—aging equipment (e.g., U.S. metalworking equipment is less than half as new as Japan and Germany, 63 percent of whose equipment is less than ten years old while only 33 percent of U.S. equipment is less than ten years old).

In table 10.5 we present the profitability and productivity of ten industries, such as instruments, machinery, non-ferrous metals. The two measures of profitability can be combined, giving equal weights to profit per dollar of sale and profit per dollar of equity. Instruments and related products, with a combined score of 26.8, is the first; motor vehicles and equipment is the tenth, with a combined score of -12.7.

Similarly, the comparable data on productivity, giving equal weight to index of productivity and capital stock per hour worked gives a score of 178.52 for instruments and related products, followed by machinery, excluding electrical, with a score of 174.72. Primary metal industries is the fifth among the five industries on which the productivity data are available.

If we take into account the fact that metalworking equipment in the United States is half as new as that in Japan and Germany, primary metal industries, which rank number seven out of ten in profitability and fifth out of five in productivity is definitely a candidate industry for robotics.

So is motor vehicles and equipment, which ranks tenth in ten in profitability and fifth in 5 in productivity as represented by capacity utilization. The additional fact that 25 percent or more of the domestic market is already lost to Japan makes motor vehicles and equipment industry definitely a candidate industry for robotics.

At the other end of the robotics candidate spectrum are two industries which do not need robotics, but can indeed afford it by virtue of profitability and productivity: Instruments and related products, which ranks first out of ten in profitabil-

ity, and ranks first out of five in productivity; and machinery, excluding electrical, which ranks second out of ten in profitability, and ranks second out of five in productivity.

We can now refine the earlier alignment of capital alter ego and capital anti-ego in robotics to identify profitability and productivity factors:

- Capital alter ego *with* long-term profitability *against* short-term profitability *with* long-term productivity *against* short-term productivity

- Capital anti-ego *against* long-term profitability *with* short-term profitability *against* long-term productivity *with* short-term productivity.

This refinement identifies the specific sets of values that have to be developed in waging the alter ego conflict in each industry. It bears repeating that the personification of the conflict is designed to specify the persons (interests) who argue vigorously for the particular viewpoint. The adversary positions tend to sharpen the issues at stake, e.g. next poor quarterly report denying the capital funds required for long-term productivity via robotics.

0.0 Conflicting Technology Concerns—Piecemeal Robotics vs. Wholesale Robotics

If the U.S. industry is engaged in a decisive combat to retake its *domestic* market from Japan first, and then acquire and extend its world market, the most significant element will be the technology that would not only equal, but also excel the Japanese technology. With Japan in the forefront of robotization, and announced national commitment to push robotization, the United States industry, with machinery half as new as the Japanese and German (in the metal-working field), indeed has a lot of catching up to do. The pace of this catching up is the issue in discussing piecemeal robotics.

0.0 Piecemeal Productivity Through Island Robotics?

At the August 1982 seminar "Robotics and the Factory of the Future," sponsored by the University of Michigan and the Michigan Chamber of Commerce, James Baker, executive vice president of the General Electric Company, suggested that it is possible to "start with islands" of robotics:

> It's possible to start with islands, a robot here and NC [Numerical control] and DNC [Direct Numerical Control], controllable machine there, and then go to linked islands and eventually to *large factory systems*. And then, if you can really need or can justify it, to an *almost totally automated factory*.
>
> The tendency sometimes is to consider automation a luxury, an indulgence you allow yourself when the numbers

254

Table 10.5
Industry Ranking by Profitability and Productivity

	PROFITABILITY (1)	(2) Per $ Sale	(3) Per $ Equity	PRODUCTIVITY (4) (2)+(3)	(5)	(6) Index of Ind.Prodn.	(7) Capital Stock	(8) (6)+(7)	(9)	(10) Capacity Utilization
Industry										
Instruments and related products		9.3%	17.5%	26.8	Rank 1	171	7.52	178.52	Rank 1	90
Machinery, excl. electrical		6.5%	15.0%	21.5	Rank 2	163	11.72	174.72	Rank 2	
Non-ferrous metals		5.9%	15.6%	21.5	Rank 3					
Aircraft, guided missiles and parts		4.3%	16.0%	20.3	Rank 4					79
Electrical and electronic equipment		5.0%	15.1%	20.1	Rank 5					
Fabricated metal products		4.2%	13.9%	18.1	Rank 6	135	9.30	144.30	Rank 4	74
Primary metal industries		4.1%	11.7%	15.8	Rank 7	102	29.65	131.65	Rank 5	73
Stone, clay, and glass products		4.2%	10.8%	15.0	Rank 8	147	15.29	162.29	Rank 3	
Iron and steel		2.9%	9.0%	11.9	Rank 9					67
Motor vehicles and equipment		-3.4%	-9.3%	-12.7	Rank 10					

are good. Well, it's not. It's *survival insurance*, and it must be done even when things are tight—even *especially when things are tight*. It hurts to pay the price for some of this stuff when business isn't so good, but trade-offs and sacrifices have to be made before things can get better.[15] (emphasis added)

Baker identifies associations of opposing interests. Since the pro and con arguments with respect to robotics come from within the same organism, i.e., the same company, we could continue to use the personification of these opposing interests in the personages of alter ego and anti-ego of the company, one or more executives arguing vigorously for each viewpoint:

- Company alter ego *with* existing plant *against* new plant.

We may consider this viewpoint less progressive than the opposite viewpoint. Attributing it to company anti-ego, instead of to the company alter ego, we have the following CONCOL:

- Company alter ego *with* new plant *against* existing plant
- Company anti-ego *against* new plant *with* existing plant.

We have already seen that in metalworking, the U.S. has equipment half as new as that in Japan and Germany. A piecemeal introduction of automation, in the form of an early vintage piece like NC, is quite unlikely to get us far. We should note that to Japan, NC is the "first stage of automation," a stage that she has long since passed. If the U.S., starting much later, thinks in terms of an NC here and a DNC there, it is going to find itself competing with a country where *by 1978* more than 52 percent (by value) of lathes were equipped with NC:

> *The first stage of automation* in the batch production mechanical engineering industries is supported by numerically controlled *(NC) machines*. At this stage, NC milling machines are the most important application. . . .
>
> The numerically controlled installations have been equipped with machine tools for such tasks as *turning, milling, drilling, boring*, and *grinding. Machining* centers appeared with NC machines from the outset. It is worth noting that the *rate of increase in machining centers is considerably higher* than in any other tools with the exception of lathes. . . .
>
> Of lathes produced in 1978, 21.7 percent by number were equipped with NC, 52.4 percent by value. It is probable

15. Rita R. Schrieber, *op. cit.*, 42.

that almost all machine tools will some day be equipped with NC.[16] (emphasis added)

The CONCOL players in machinery modernization are:

- Company alter ego *with* full-scale robotics *against* piecemeal automation *with* significant productivity promotion
- Company anti-ego *against* full-scale robotics *with* piecemeal automation *against* significant productivity promotion.

0.0 How Not to Robotize Piecemeal: Ohgishima Automation

An excellent example of nonpiecemeal productivity improvement is Ohgishima steel mill which in 1980 produced eighteen hundred tons per worker, *more than three times* that of Bethlehem Steel Company's modern Burns Harbor plant. Two factors of tremendous importance that made the modernization possible were: (1) preagreement on manpower utilization between management and union, which was greatly facilitated by the fact of (2) company, instead of national, unions.

0.0 Prior Agreement with Company Union

Gibney asked a department head who had previously served as a company union executive the reason for the successful transition. The answer was:

> We first had to talk to the union *years in advance of building the new Ohgishima works* and explain exactly what we had in mind. The jobs here would be far fewer, of course, but they were good jobs. We were *eliminating the dirty jobs*. No one wants a dirty job if he can avoid it. Union members knew that they would either be *trained* for the new jobs or *transferred* to other branches where new and better jobs awaited them. We *planned for this far in advance*.[17] (emphasis added)

0.0 New Skills, New Perspectives

Training and transferring can be looked upon as part of the capital cost of building and operating the new Ohgishima plant. People were treated as being more important than the machinery. Well before the new plant building was erected, the training and/or transferring was worked out in great detail. No piecemeal automation here, no token NC or DNC here.

16. Committee on Computer-Aided Manufacturing, "State of the Art in Japan," in "Computer-Aided Manufacturing: An International Comparison" (Washington, D.C.: National Academy Press, 1981), 3–4.

17. Gibney, *op. cit.*, 94, 148–49.

Robotics/Artificial Intelligence/Productivity

More than a shiny new plant, Ohgishima automation called for a new perspective. When the new plant would open, the much fewer workers who would work there were going to be part of a new wave of production. As such, they had to be thoroughly inculcated with the new outlook, attitude, and eagerness to make the new automated plant succeed.

It is this transformation of perspectives and skills that is ignored in a piecemeal approach to robotics. A token NC or DNC would probably have as much chance of success in improving America's production and productivity as the installation of an automatic counter on the wheel of a buggy has in competing with the automobile.

The central role of training is emphasized by Thomas Mathues, vice president of General Motors:

> Robotics is only one means we are using to increase productivity and quality, but it's one of the most important. . . . We anticipate that these R & D advances and others to come will lead to robots that use sensors and end effectors for *adaptive control for parts acquisition and inspection*. These robots will be fully integrated in the manufacturing operations with effective people-to-people and people-to-robot interfaces. They also will be readily programmed using advanced offline support systems. . . .
>
> The training aspects of robotics can't be overstated. The job of *training and retraining is immense and [is] a real bottleneck in the race to automate*. [G.] Paul Russo [director of manufacturing at Chrysler] agrees: "We'd be moving much faster in my own corporation if we had the people," he said. "We've got the training programs for technicians and skilled tradesmen, manufacturing engineers, design and process people, but we're not training people fast enough. *The time [it takes] to train them is longer than we anticipated*, too, and they seem to come back for refresher courses."
>
> "At Chrysler we have some 80 robots compared to 80,000 people. That's a very small percentage that have to be retrained. And this has occurred over several years. So the replacement of an employee can very easily be accommodated by not hiring someone for a job that comes open through attrition or any other way. Although it seems big in everyone's mind, I believe the *retraining*

problem is miniscule compared to the training problem.[18]
(emphasis added)

0.0 Keep the Job, Lose the Plant

It may well be that the industry-wide unions, such as the United Auto Workers (UAW), would find that it is easier to let plants close altogether than have the union members trained, retrained, and relocated. The "success" from the point of view of UAW may at best be a pyrrhic victory to the unemployed worker. From the point of view of management, which has to beg and borrow sizeable capital investment funds that would take several years to show dividends, additional sunk costs of worker training, retraining, and relocation may be too high a risk with the prospects of losing the newly trained, retrained, relocated worker to the competition.

In Japan, the unions are not industry-wide, but company-wide. Each company's unions have access to the company books, making them more a collaborative management partner than an adversary. Doubtless, there are the rituals of voluble bargaining, demand and counter-demands, charges, and counter-charges, all exchanged to let off steam and to demonstrate the vociferous protection of members' interests by the union. The union knows, however, that it is their company just as much as the management's and that the workers stay with the company for their entire working life. With the life-time employment commitment to each other, the spring bargaining, and even strikes, are engaged in as responsible partners willing to sink their differences to make the company survive and grow, so that both can reap the benefits.

The issues which militate against wholesale robotics in the United States are to be found in several areas: (1) required dramatic change in perspective and skills, (2) training/retraining/relocation costs, which should be considered as part and parcel of (3) significant capital assets outlay, (4) adversary relationship between labor and management, and (5) industry-wide unions.

The CONCOL players are:

- Company alter ego *with* capital outlay for robotics *with* capital outlay for training/retraining/relocation *with* job changes *with* joint planning with labor
- Company anti-ego *against* capital outlay for robotics *against* capital outlay for training/retraining/relocation *against* job changes *against* joint planning with labor.

18. Schrieber, *op. cit.*, 40–42.

PART V

Concomitant Coalitions Between the United States and Japan

11

The Context of Concomitant Coalition Process

OVERVIEW: Balancing the opposing concerns of labor, capital, and technology *within* the U.S. is critical to the success of balancing the opposing concerns of high-technology imports followed by high-tech products exports in CONCOLs *between* the U.S. and trade partners such as Japan, Taiwan, and Korea. The CONCOLs begin with the starting negotiating positions of each player. Having identified the various *players* in the twenty-four major CONCOLs within Japan and the three major groups of CONCOLs within the U.S., we turn to the *process* of CONCOL, the three critical elements of which are: *areas of bargaining (Why?), measures of outcome (What?),* and *changes by bargaining (How?).*

The CONCOL process is illustrated by applying it to the Kennedy-Khrushchev confrontation on the Cuban Missile Crisis in which the initial positions contained nonnegotiable and negotiable areas. The key to the solution lay in first identifying which *areas* held vital interests to each party and which aspects were nonnegotiable.

To measure the negotiability or nonnegotiability of the issue to each party, it is necessary to develop an index of its imperativeness based on the ratio of (need/want) represented by that issue. A *need* is mandatory for the organism to fulfill its primary functions. A *want* is nonmandatory to the organism in fulfilling its primary functions. The need for transportation, for instance, is met by walking, taking the bus, taking the subway, or driving the personal car. If the personal car is used, strictly from the point of view of fulfilling the transportation need, a Volkswagen Beetle or Lincoln Continental or a number of alternatives in between would do, but only one can best meet the "want" of the particular party.

An issue, event, or activity that is all want and no need is represented as 1% need (to avoid dividing zero and dividing by zero), and 99% want, giving rise to a (need/want) ratio of (1%/99%), or 0.01. At the other extreme, of all need and no want, similarly, is (99%/1%), or 99. This ratio is called the N-Index. In the former instance, the N-Index can be expressed as 10^{-2} in the latter, 10^2. In the former instance, the N-Index exponent is negative; in the latter, it is positive. In

between the two is *N*-zero. In the Kennedy-Khrushchev CONCOL, the process was to move each party from an initial *N*-Positive position to an *N*-Zero position.

Turning from international politics to international trade, we identify: *macroelements, minielements,* and *microelements.* A *quantitative* initial desirable outcome is identified with respect to the seven macroelements, four minielements, and three microelements as providing the context of the CONCOL process with respect to high-technology and high-tech products trade in general, and in particular, Fifth Generation Computer (5G) and associated products.

0.0 CONCOL Outcome: Negotiated Equivalence

Given that, in international trade, as elsewhere, no one participant (party, player, interest) can acquire the entire outcome, prudence dictates that even the strongest player be not unwilling to accept reasonable compromises. Unless one player controls all the variables affecting the outcome, every player controlling each variable will normally seek to achieve the maximum value for himself or herself. And since everyone cannot achieve the maximum, the players can either fight each other to death or agree among themselves as to how best to split the outcome. When one cooperates with one's adversary, he or she is assuming a position opposite to that of relentlessly pursuing his or her maximum gain. Insofar as the cooperation is for long-term survival and success of one's endeavours, the association with and against the same party or parties in the same game is a concomitant coalition (CONCOL):

- Player alter ego *with* the competition *against* short-term success *with* long-term survival and success
- Player anti-ego *against* the competition *with* short-term success *against* long-term survival and success.

In chapters 9 and 10, we have identified the main players in the CONCOL in Japan and in the United States in the Robotics Revolution. We saw in chapter 9 that twenty-four major CONCOLs identify the significant opposing interests whose skillful balancing accounts for the visible success of Japanese productivity. We also saw in chapter 10 that CONCOLs representing labor concerns, capital concerns, and technology concerns have to be reckoned with in the coming Robotics Revolution in the United States. With the *players* identified, we now turn to the *process* of CONCOL, the three critical elements of which are: *areas of bargaining (Why?), measures of outcome (What?),* and *changes by bargaining (How?).*

0.0 Areas of Bargaining (Why?)

The inherent, simultaneous opposition of objectives is the essence of CONCOL. Very simply, the United States and Japan, seeking to maximize their share of the

world market in any product and/or process, e.g., robotics, microprocessors, etc., constitute the essence of CONCOL. Each party to a CONCOL: (1) must have inherently opposing interests that each pursues simultaneously, (2) must be able to pursue the opposing objective(s) with the same party or parties at the same time, and (3) must have measurable objectives, the probability of outcome of which must permit decisive comparisons to permit choices.

To identify measures of objectives, it is essential to start with areas of interest to the bargaining by the members of the concomitant coalition. Bargaining is the process of offer and counter-offer by the parties to move the outcome closer to the objective of each party. Clearly, before there can be bargaining, there must be perceived areas where both parties must feel it important to accomplish objectives; otherwise, they would have no reason to bother trying to influence the outcome.

To facilitate discussion of the crucial elements of the CONCOL bargaining process, let us use a historic example, the Cuban Missile Crisis.

It is interesting to note that Khrushchev uses terminology about Kennedy that emphasizes the simultaneously opposite roles that they played:

> I'd like to say a few words about the way Kennedy conducted his side of the Vienna talks. We were sitting in a room with only our interpreters, Rusk and Gromyko. . . . John Kennedy and I met man to man, as the two principal representatives of our countries. He felt perfectly confident to answer questions and make points of his own. This was to his credit, and he rose in my estimation at once. *He was, so to speak, both my partner and my adversary.* Insofar as we held different positions, he was my *adversary*; but insofar as we were negotiating with each other and exchanging views, he was my *partner* whom I treated with great respect.[1] (emphasis added)

We may paraphrase Khrushchev as follows:

- Khrushchev *with* Kennedy on coexistence
- Khrushchev *against* Kennedy on status quo maintenance:

> Kennedy wanted to maintain the status quo in the world. I was also in favor of the status quo, and still am, but we differed in our understanding of what this term meant. For us "maintaining the status quo" meant agreeing not to violate the borders that came into existence after World War II—and especially not to violate them by means of war. Kennedy, however, had in mind the inviolability of

1. Strobe Talbott, ed. and trans., *Khrushchev Remembers* (Boston: Little, Brown, 1974), 497 –98.

borders *plus the enforced preservation of a country's internal social and political system.* In other words, he wanted countries with capitalist systems to remain capitalist, and he wanted us to agree to a guarantee to that effect.

This was absolutely unacceptable. At that time many people still lived under colonialism. Did he really expect us to help the colonialists continue their oppression of their colonies?[2] (emphasis added)

In light of Khrushchev's amplification, we can sharpen the agreement, as well as the disagreement, between Khrushchev and Kennedy. "Coexistence" refers to the United States and the Soviet Union pursuing their present courses. In other words, it is status quo as far as superpowers are concerned. When it comes to status quo in countries other than the United States and USSR, however, Kennedy wants them to be left alone, but Khrushchev wants to have the freedom to engage in "wars of liberation." Thus we could restate the concomitant coalition position as follows:

- Khrushchev *with* Kennedy on status quo of superpowers
- Khrushchev *against* Kennedy on status quo of nonsuperpowers.

Within the framework of this concomitant coalition, negotiations are possible. Khrushchev could not affect the status quo in *all* the nonsuperpowers at the same time. He would, therefore, be necessarily limited to confining his actions to upsetting the status quo in *some* of the non superpowers. If Kennedy had a special interest in some of these countries, Khrushchev could agree not to rock the boat in those countries for the time being. In return, Khrushchev could demand maintaining the profile he desired in the other nonsuperpower countries. Or, he could demand contributions to the maintenance of status quo of the superpowers themselves.

In the Khrushchev-Kennedy CONCOL, the status quo of the superpowers was considered inviolate. In other words, the economic, social, political, and military aspects of the superpower profile could not be affected by the other superpower. While it is fully recognized that each power tries to influence the other, the maintenance of status quo requires that no drastic change in the existing economic, social, political, and military aspects be tolerated.

What areas of the nonsuperpowers profile is negotiable?

Changes in the current profile of some of the countries could be agreeable; but not of some other countries. The *first measure*, therefore, is the number of nonsuperpower countries that could be affected. Even among the countries whose profile could be affected, changes in certain aspects of the country could be toler-

2. *Ibid.*, 495–96.

ated, but not changes in the military aspects; or changes in the political might be tolerated, but not in the military, etc.

The bargaining between the CONCOL members becomes the give-and-take on the basis of the evaluation by each party of the significance of the projected changes. The projected changes are by no means certain; therefore, the evaluation has to be in terms of probabilities—probabilities of actions and probabilities of results. The question of credibility of the promises by the other members is crucial.

We could reconstruct the Khrushchev-Kennedy CONCOL in terms of negotiable and nonnegotiable areas:

- **Nonnegotiable:** Superpower Profile—economic and social; political and military aspects
  Nonsuperpowers Profile—military aspects
- **Negotiable:** Nonsuperpowers Profile—economic and social aspects
- **Quasi-Negotiable:** Nonsuperpowers Profile—political aspects

0.0 Measures of Outcome (What?)

It is one thing to identify *areas* of bargaining; it is quite another to assess acceptable outcomes. For instance, what does it mean to say that the military aspects of nonsuperpowers is not negotiable? Does it mean that the country in question cannot receive weapons supplies from external sources? Does it mean that the country in question cannot have technical advisers from outside? Does it mean that the training abroad of the present and/or potential leaders of the country in question is unacceptable?

It is in the very nature of bargaining that the initial positions taken by the parties to the bargaining are subject to change. The change will take place in a reasonably equitable a manner, whatever is given up by one side being approximately equal to whatever it has gained.

Here's the rub. What is the equitable measure to the other side? What does it consider important? This cannot be answered without referring to the value system of the other side.

The Cuban Missile Crisis, in retrospect, is an example of miscalculation of the outcome by the two parties involved. According to the *Last Testament*, Khrushchev believed that the United States, unhappy over the Bay of Pigs, would seek to overthrow Castro by bringing U.S. military in support of "counterrevolutionary forces" in Cuba. Because of the five-thousand-mile distance, the Soviet Union would not be able to assist Cuba in that eventuality. That Cuba should succeed in her "democratic experiment" was of paramount concern to Khrushchev as part of his philosophy of "wars of liberation," which would set countries free from the yoke of the "colonialists." As a "defensive measure," therefore, simply to protect Cuba from "U.S. aggression," Khrushchev introduced Soviet missiles into Cuba. From Khrushchev's point of view, the introduction of Soviet missiles into Cuba was simply to maintain the status quo in the *economic and social aspects* of

a nonsuperpower—aspects which were clearly negotiable, as we saw at the end of the previous section.

Kennedy, however, saw it as something entirely different.

He had no plans to invade Cuba or to support an indigenous movement to overthrow Castro. Having accepted full blame for the Bay of Pigs fiasco, he would not rush into another Bay of Pigs. Suddenly, he is faced with missiles pointed at the heartland of America. At a time when the accuracy of missiles was far from established, missiles located 5,000 miles away in the U.S.S.R. and U.S.A. would be approximately equal in their effectiveness or ineffectiveness. But when the U.S. missiles are 5,000 miles away and the U.S.S.R. missiles are only 90 miles away, that does tip the military advantage decisively in favor of the U.S.S.R. From Kennedy's point of view, therefore, the introduction of Soviet missiles into Cuba was *not* to maintain the status quo in the economic and social aspects of a nonsuperpower, but, instead, to dramatically change the status quo of the *military* aspects of the superpower profile, which, as we saw in the last section, was nonnegotiable.

To *measure* the negotiability and nonnegotiability of the issue to each party, it is necessary to develop an index of each issue's imperativeness.

We do that by making a distinction between "needs" and "wants." The need for transportation, for instance, is met by walking, taking the bus, taking the subway, or driving the personal car. If the need is to be met by driving the personal car, strictly from the point of view of fulfilling the need for transportation, a vehicle as functional as the Volkswagen Beetle or as luxurious as the Lincoln Continental could do the job. While the "need" for personal transportation is met by either, or a number of alternatives in between, only one can meet the "want."

A *need* is mandatory for the organism to fulfill its primary functions. A *want* is nonmandatory to the organism in fulfilling its primary functions.

It should be recognized that the "want" of an earlier day may well become the "need" of a later day. Further, the assertion by one party to a CONCOL that something is a "need" could be a bargaining posture. To make an operational measure out of the need-want combinations, it is fitting that a ratio incorporating admissible combinations be developed. Since the extremes, i.e. all want and no need, or all need and no want—100% need and 0% want, or 100% want and 0% need—would make the ratio of (need/want) indeterminate, we could use instead 1% instead of 0%. The extreme ratios would thus be:

- all want and no need = 99% want and 1% need;(need/want) = 1%/99% = 1/99 = 0.01
- all need and no want = 1% want and 99% need;(need/want) = 99%/1% = 99/1 = 99.00

We can call the ratio of (need/want) the Need Ratio, or the N-Index. The N-Index can range from 0.01 to 99, or approximately 10^{-2} to 10^2. When need

= want, the ratio is 1, or 10^0. Thus, the *exponent* of the N-Index can be negative, zero, or positive, making the situations N-negative, N-zero, or N-positive.

We stated in the last section that the superpower profile was nonnegotiable; which means that for both Kennedy and Khrushchev, the N-Index is positive, the maintenance of superpower profile unchanged being an imperative "need," making the numerator of the (need/want) ratio definitely higher than the denominator, resulting in a positive ratio, expressed as a positive exponent. We also said that the military aspects of nonsuperpower profiles were also nonnegotiable. The two N-Positive positions led to a direct encounter. We can present the *initial* positions of the encounter as follows:

- N-positive Khrushchev *with* N-positive Kennedy on superpower profile status quo
- N-positive Khrushchev *against* N-positive Kennedy on superpower profile alteration
- N-positive Khrushchev *with* N-zero Kennedy on nonsuperpower economic and social profile alteration
- N-positive Khrushchev *against* N-positive Kennedy on nonsuperpower military profile alteration

0.0 Changes by Bargaining (How?)

Accommodations would be easier on N-zero positions than on N-positive positions. Taking the statements of Khrushchev in the *Last Testament* at face value, the reason for his introducing the missiles into Cuba was purely to ensure that the Cuban experiment in democratic living would be permitted to be conducted without foreign intervention. In other words, he had no interest in the military aspects of the nonsuperpower profile. Thus we could consider Khrushchev to be N-zero on superpower military profile alteration, but N-positive on superpower economic and social profile alteration. At the same time, Kennedy considered the particular nonsuperpower military profile alteration to be quite significant. Kennedy, therefore, would be N-positive on nonsuperpower military profile alteration. We can present the situation as follows:

- N-positive Khrushchev *against* N-zero Kennedy on nonsuperpower economic and social profile alteration
- N-zero Khrushchev *with* N-positive Kennedy on nonsuperpower military profile alteration.

Kennedy interpreted Khrushchev's claim to *nonsuperpower economic and social* profile alteration to be in fact *superpower military* profile alteration. Both would agree on the issue of superpower military profile *status quo*; therefore, the Kennedy object of the Cuban Missile Crisis negotiations was to persuade Khrushchev that the superpower military profile was in fact altered; that it had to

be restored. At the same time, taking *Last Testament* at face value, Kennedy had to be persuaded that the nonsuperpower economic and social profile was threatened; the threat had to be removed. We can understand the purpose of the bargaining to be that of letting each side know what the other side is *N*-positive about:

- N-positive Kennedy *against* N-zero Khrushchev on superpower military profile alteration
- N-positive Khrushchev *against* N-zero Kennedy on nonsuperpower economic and social profile alteration.

The solution to the Cuban Missile Crisis could be interpreted as the progress toward *N*-zero positions in the place of *N*-positive positions. Since the U.S. did not intend to invade Cuba or to support an anti-Cuban uprising from within, the U.S. was *N*-zero with respect to the economic and social profile alteration of Cuba. Similarly, strictly on the basis of *Last Testament*, the U.S.S.R. was *N*-zero with respect to superpower military profile. Withdrawal of Soviet missiles from Cuba on the U.S. assurance of nonintervention in Cuba would therefore provide a solution in the realm of *N*-zero.

0.0 Starting Negotiating Positions of CONCOL Players—Macroelements

To appropriate the productivity promises of robotics and artificial intelligence, both the U.S. and Japan will find themselves cooperating and competing simultaneously. The cooperation will be in the realm of high-technology; the competition will be in the area of high-technology products that have to find a place in the world market.

It is appropriate to separate the CONCOL considerations into *macro* and *micro*. The macroelements deal with the bottom line, so to speak, of the trade game, viz., the balance of trade and its components—exports and imports. The value of world exports rose from $58 billion in U.S. currency in 1950 to $1,868 billion in 1980, registering an average annual growth rate of 12.3 percent in nominal terms and 6.7 percent in real terms.[3]

0.0 Macroelement 1: Share of World Exports

How much of the growth in world exports has the particular country shared?

Taking a fifteen-year span, the U.S. exported $30.4 billion in 1966, which was 21.9 percent of the world exports. In 1981, the U.S. exported $233.7 billion, which was 19.2 percent of the world exports.

3. Unless otherwise specified, the trade statistics cited come from the International Monetary Fund, *Supplement on Trade Statistics* (Washington, D.C.: 1982).

Japanese exports of $9.8 billion in 1966 (7.1 percent of world exports) rose to $151.5 billion (12.4 percent of world exports).

A specific share of world exports, e.g., 20 percent for the U.S., and 10 percent for Japan, could be the initial desired outcome.

What should be the nature of the *N*-Index?

Should each nation insist that the chosen percentage be achieved, that it is *nonnegotiable?* Perhaps not. For one thing, the particular share is the result of transactions world-wide, several of which could have a significant impact on shifting the percentage figure one way or the other. Further, if a country were to insist on a particular share of world exports, with whom will it negotiate to achieve that result? While major trading partners would be plausible parties to the negotiation, the final determination of a country's share of *world* exports will be beyond the powers of the most powerful of the trading partners themselves.

The "share of world exports" appears to be an *N*-zero index; so also are the other macroelements.

0.0 Macroelement 2: Share of World Imports

In 1966, the U.S. imported $27.8 billion, which was 14.3 percent of world imports. In 1981, the U.S. imported $273.3 billion, which was also 14.3 percent of world imports.

In 1966, Japan imported $1.1 billion, 0.6 percent of world imports. In 1981, Japan imported $142.9 billion, 7.5 percent of world imports.

A specific share of world imports, e.g., 14 percent for the U.S. and 7 percent for Japan, could be the initial desired outcome.

0.0 Macroelement 3: Balance of Trade

The world balance of trade in 1961 was US$-5.98 billion. Japan had a negative balance of trade of US$-1.58 billion, accounting for 26.4 percent of the world deficit. The U.S. enjoyed a positive balance of trade of $5.08 billion.

Twenty years later, in 1981, the world balance was still negative at US$-72.48 billion. The U.S. had a negative balance of $39.61 billion, accounting for 54.6 percent of the world deficit. This time, Japan enjoyed a positive balance of trade of $8.63 billion.

If we take the average of the U.S. balance of trade in 1971 and 1981, i.e., -4.60 and -39.61 respectively, as the initial desired outcome, we have -22.1 billion. For Japan, the corresponding figure is the average of $+4.32$ and $+8.63$, or 6.5 billion.

0.0 Macroelement 4: Terms of Trade Index

The terms of trade index is the ratio of export unit value to import unit value.

In 1951, the U.S. terms of trade index was 100.2, Japan's 132.1. Thirty years later, they were, respectively, 84.9 and 78.5.

The average of 1971 and 1981 values would yield 102.3 for the U.S., and 108.8 for Japan. The ideal value of 100.0 could be a desired initial outcome.

0.0 Macroelement 5: Ratio of Exports to GDP
In 1966, U.S. exports represented 4.1% of the Gross Domestic Product; in 1981, they rose to 8.1 percent.

In 1966, Japanese exports represented 9.3 percent of Gross Domestic Product; in 1981, they rose to 13.5 percent.

The average of 1971 and 1981 values would yield 6.1 percent for the U.S. and 12.0 percent for Japan for the desired initial outcome.

0.0 Macroelement 6: Ratio of Imports to GDP
In 1966, U.S. imports represented 3.7 percent of Gross Domestic Product; in 1981, they rose to 9.5 percent.

In 1966, Japanese imports represented 9.0 percent of Gross Domestic Product; in 1981 they rose to 12.6 percent.

The average of 1971 and 1981 values would yield 7.0 percent for the U.S. and 10.6 percent for Japan for the desired initial outcome.

0.0 Macroelement 7: Weeks of Reserve Coverage
Merchandise imports have been related to the total reserves of the country, comprising gold and Standard Drawing Rights (SDR). For the world as a whole, reserve coverage declined from 41 weeks of imports in 1950 to just under 12 weeks of imports in 1981. The reserve coverage for the United States fell from a high of 178 weeks of imports in 1949 to about 5 weeks in 1981. The average of 1971 and 1981 values yields 9 weeks for the U.S. desired initial outcome.

0.0 Starting Negotiating Positions of CONCOL
Players—Minielements
While the N-zero Index values of macroelements indicate the bottom line, so to speak, of the high-technology activities culminating in high-technology products trade, the minielements are the ones which bear on the CONCOL process directly.

0.0 Minielement 1: Country Group Share of Exports
In table 11.1, we present the direction of export flows from selected countries as percent of the total. In 1966, industrial countries accounted for 72.1% of the world exports, oil-exporting countries 3.2%, nonoil developing countries 22.4%, and other countries 2.2%. Fifteen years later, in 1981, the share of the industrial countries declined to 66.4%, or a drop of 5.7 percentage points. The drop in the share of the industrial countries was picked up by oil-exporting countries, which increased their share by 4.8 percentage points, and other countries, which increased their share by 1.1 percentage points, the share of the nonoil developing countries remaining unchanged.

The Context of Concomitant Coalition CONCOL Process

Table 11.1
Direction of Export Flows (% of Total)

		1966	1971	1976	1981	1981–1966
World						
Exports	BIL $	138.7	247.7	633.3	1,219.8	1,081.1
Industrial countries	%	72.1	72.9	69.2	66.4	− 5.7
Oil-exporting countries	%	3.2	3.4	7.0	8.0	4.8
Nonoil developing countries	%	22.4	21.2	20.0	22.4	0.0
Other countries	%	2.2	2.4	3.8	3.3	1.1
U.S.						
Exports	BIL $	30.4	44.2	115.4	233.7	203.3
Industrial countries	%	62.6	65.3	58.5	55.5	− 7.1
Oil-exporting countries	%	4.6	5.0	10.5	8.9	4.3
Nonoil developing countries	%	31.5	27.9	26.1	31.8	0.3
Other countries	%	1.3	1.9	4.8	3.9	2.6
Ratio of exports to GDP	%	4.1	4.1	6.8	8.1	4.0
Japan						
Exports	BIL $	9.8	24.1	67.3	151.5	141.7
Industrial countries	%	49.2	51.5	44.6	46.4	− 2.8
Oil-exporting countries	%	4.9	5.5	13.7	15.1	10.2
Nonoil developing countries	%	37.2	32.5	29.8	31.1	− 6.1
Other countries	%	8.6	10.5	11.9	7.4	− 1.2
Ratio of exports to GDP	%	9.3	10.4	12.0	13.5	4.2
Germany						
Exports	BIL $	20.2	39.1	102.2	176.1	155.9
Industrial countries	%	77.2	77.7	71.3	71.7	− 5.5
Oil-exporting countries	%	3.2	2.9	8.0	8.7	5.2
Nonoil developing countries	%	16.8	16.0	15.0	15.5	− 1.3
Other countries	%	2.8	3.4	5.7	4.1	1.3
Ratio of exports to GDP	%	16.5	18.0	22.9	25.6	9.1
Nonoil developing countries—Asia						
Exports	BIL $	10.5	15.3	51.2	129.6	119.1
Industrial countries	%	61.8	59.7	62.6	54.9	− 6.9
Oil-exporting countries	%	4.0	4.0	7.5	8.5	4.5
Nonoil developing countries	%	28.6	30.2	25.6	32.5	3.9
Other countries	%	5.5	6.1	4.4	4.0	− 1.5
China PRC						
Exports	BIL $				21.6	
Industrial countries	%	45.0	38.3	44.3	49.7	+ 4.7
Oil-exporting countries	%	6.8	7.3	10.0	6.2	− 0.6

273

Table 11.1 (*continued*)

		1966	1971	1976	1981	1981–1966
Nonoil developing countries	%	48.2	54.4	45.7	44.1	− 4.1
Other countries	%	-	-	-	-	-
Hong Kong						
Exports	BIL $	1.3	2.9	8.5	21.8	20.5
Industrial countries	%	67.0	74.6	73.5	59.5	− 7.5
Oil-exporting countries	%	8.1	4.8	6.8	9.8	1.7
Nonoil developing countries	%	23.5	17.3	16.7	27.4	3.9
Other countries	%	1.4	3.2	3.0	3.3	1.9
Ratio of exports to GDP	%	67.4	78.5	88.0	90.3	22.9
Korea						
Exports	BIL $	0.2	1.1	7.7	21.3	21.1
Industrial countries	%	81.3	85.1	78.5	62.8	− 18.5
Oil-exporting countries	%	0.8	3.1	9.3	13.2	12.4
Nonoil developing countries	%	17.0	10.7	11.0	18.2	1.2
Other countries	%	0.9	1.1	1.2	5.8	4.9
Ratio of exports to GDP	%	6.6	11.3	28.8	32.3	25.7
Singapore						
Exports	BIL $	1.1	1.8	6.6	21.0	19.9
Industrial countries	%	48.3	40.0	48.9	40.7	− 7.6
Oil-exporting countries	%	2.2	1.7	4.2	7.0	4.8
Nonoil developing countries	%	48.1	54.4	44.2	49.7	1.6
Other countries	%	1.4	3.9	2.6	2.6	1.2
Ratio of exports to GDP	%	101.3	78.7	111.6	162.4	61.1

Since nonoil developing countries listed in table 11.2, such as Korea, want to import high-technology from the U.S. and Japan in order to export high-tech products to them, the two country groups of particular importance are *industrial countries*, and *nonoil developing countries*. The 1966 and 1981 figures of selected countries, as well as the average of 1976 and 1981, are as follows:

	Industrial Countries			Non-oil Developing Countries		
	1966	1981	1976/81 average	1966	1981	1976/81 average
U.S.	62.6	55.5	57.0%	31.5	31.8	29.0%
Japan	49.2	46.4	45.5%	37.2	31.1	30.5%

Germany	77.2	71.7	71.5%	16.8	15.5	15.2%
Hong Kong	67.0	59.5	66.5%	23.5	27.4	22.0%
Korea	81.3	62.8	70.6%	17.0	18.2	14.6%
Singapore	48.3	40.7	44.8%	48.1	49.7	46.9%

What should be the nature of the N-Index?

Perhaps the individual country's export share, not of the industrial world as a whole, but by each major trading group of countries, is less nonnegotiable than a world-wide figure. To reflect the more-need-and-less-want characterization, the need/want ratio would be greater than one or the N-Index would be positive.

0.0 Minielement 2: Country Group Share of Imports

A similar set of initial desired outcomes can be identified from table 11.3 as follows:

	Industrial Countries			Nonoil Developing Countries		
	1966	1981	1976/81 average	1966	1981	1976/81 average
U.S.	66.7	52.4	52.5%	26.0	25.6	23.9%
Japan	49.8	34.6	36.7%	30.0	22.2	22.2%
Germany	72.9	72.2	71.6%	17.0	12.8	13.3%
Hong Kong	52.7	50.4	51.3%	42.4	40.6	39.7%
Korea	85.7	62.3	66.5%	10.0	9.0	8.1%
Singapore	58.9	46.2	45.7%	26.5	25.9	27.6%

0.0 Minielement 3: Balance of Trade

At the mini level, the most significant element of the balance of trade is the balance of trade of the nonoil developing countries with the industrial countries; for it is *from* them that the nonoil developing countries have to import high-technology, and it is *to* them that the nonoil developing countries have to export high-technology products. The importance of the nonoil developing countries to the industrial countries would be reflected in the pattern of trade. The oil crisis deflected the scales in the *opposite* direction, making the nonoil developing countries less important to the industrial countries, and making the oil-exporting countries much more important:

> The industrial countries allocated 16.7 percent of their export earnings to purchase imports from nonoil developing countries in 1980, *half the 32.4* percent of 1950.
>
> The *steady decline* in the importance of the nonoil developing countries in the imports of the industrial countries

Table 11.2
Trade Supplement Guide to the Coverage of Area and World Series

Country Code	Country	Exports	Imports	Export Volume	Import Volume	Export Unit Value	Import Unit Value	Terms of Trade
001	WORLD							
110	*Industrial Countries*							
111	United States	x	x	x	x	x	x	D
156	Canada	x	x	x	x	x	x	D
193	Australia	x	x	x	x	x	x	D
158	Japan	x	x	x	x	x	x	D
196	New Zealand	x	x	x	x	x	x	D
122	Austria	x	x	x	x	x	x	D
124	Belgium	x	x	x	x	x	x	D
128	Denmark	x	x	x	x	x	x	D
172	Finland	x	x	x	x	x	x	D
132	France	x	x	x	x	x	x	D
134	Germany	x	x	x	x	x	x	D
176	Iceland	x	x	x	x	x	x	D
178	Ireland	x	x	x	x	x	x	D
136	Italy	x	x	x	x	x	x	D
652	Ghana	x	x					
656	Guinea	x	x					
654	Guinea-Bissau	x	x					
662	Ivory Coast	x	x	D	D	x	x	D
664	Kenya	x	x	x	x	x	x	D
666	Liberia	x	x	x	D	x	x	D
674	Madagascar	x	x					
676	Malawi	x	x	x	x	x	x	D
678	Mali	x	x					
682	Mauritania	x	x					
684	Mauritius	x	x	x	x	x	x	D
686	Morocco	x	x	x	x	x	x	D
688	Mozambique	x	x					
692	Niger	x	x					
696	Reunion	x	x					
714	Rwanda	x	x	x	F	F	D	D
716	São Tomé & Principe	x	x					
722	Senegal	x	x	D	D	x	x	D
718	Seychelles	x	x					
724	Sierra Leone	x	x	x	F	F		
170	*Europe*							
423	Cyprus	x	x	x	x	x	x	D
816	Faeroe Islands	x	x					
174	Greece	x	x	x	x	x	x	D
944	Hungary	x	x	x	x	x	x	D
181	Malta	x	x	x	x	x	x	D
182	Portugal	x	x	x	x	x		D
968	Romania	x	x					
186	Turkey	x	x	x	x	x	x	D
188	Yugoslavia	x	x	x	x	x	x	D
405	*Middle East*							
419	Bahrain	x	x					
469	Egypt	x	x					
436	Israel	x	x	x	x	x	x	D
439	Jordan	x	x	x	x	x	x	D
446	Lebanon	x	x	x				

138 Netherlands	x	x	x	x	x	x	D
142 Norway	x	x	x	x	x	x	D
184 Spain	x	x	x	x	x	x	D
144 Sweden	x	x	x	x	x	x	D
146 Switzerland	x	x	x	x	x	x	D
112 United Kingdom	x	x	x	x	x	x	D
200 *Developing Countries*							
999 *Oil Exporting*							
612 Algeria	x	x					
536 Indonesia	x	x	x	D	x	D	D
429 Iran	x	x	D	D	F	D	D
433 Iraq	x	x	D	D	F	D	D
443 Kuwait	x	x	D	D	F	D	D
672 Libya	x	x	D	D	F	D	D
694 Nigeria	x	x	F		F	D	D
449 Oman	x	x	D	D	F	D	D
453 Qatar	x	x	D		F		D
456 Saudi Arabia	x	x	D	D	F	D	D
466 United Arab Emirates	x	x	D	D	F	D	D
299 Venezuela	x	x	D	D	F	D	D
201 *Nonoil Developing Countries*							
605 *Africa*							
726 Somalia	x	x	x				
199 South Africa	x	x	x	x	x	x	D
732 Sudan	x	x					
738 Tanzania	x	x	D	D	x	x	D
742 Togo	x	x	F	D	x	x	D
744 Tunisia	x	x	x	x	x	x	D
746 Uganda	x	x					
748 Upper Volta	x	x	D	D	x	x	D
636 Zaire	x	x					
754 Sambia	x	x	x	D	x	x	D
698 Zimbabwe	x	x					
505 *Asia*							
512 Afganistan	x	x					
859 Samoa, American	x	x					
513 Bangladesh	x	x					
516 Brunei	x	x					
518 Burma	x	x	F		F	D	D
924 China, P. Rep. of	x	x					
819 Fiji	x	x	D	D	x	D	D
887 French Polynesia	x	x					
532 Hong Kong	x	x	x	x			
534 India	x	x	x	x	x	x	D
522 Kampuchea, Dem.	x	x					
542 Korea	x	x	x	x	x	x	D
544 Lao P. D. Rep.	x	x					
546 Macao	x	x					
548 Malaysia	x	x	F	D	F	x	D
463 Syrian Arab Rep.	x	x	x	x	x	x	D
473 Yemen Arab Rep.	x	x					
459 Yemen, P.D. Rep.	x	x					
205 *Western Hemisphere*							
311 Antigua and Barbuda	x	x					
213 Argentina	x	x					
316 Barbados	x	x					
313 Bahamas	x	x					
339 Belize	x	x					
319 Bermuda	x	x					
218 Bolivia	x	x	F	D	F	D	D
223 Brazil	x	x	x	x	x	x	D
228 Chile	x	x					
233 Colombia	x	x	D	D	x	x	D
238 Costa Rica	x	x					
321 Dominica	x	x					
243 Dominican Republic	x	x	D	D	F	D	D
248 Ecuador	x	x	D	D	F	D	D
253 El Salvador	x	x	x	x	x	x	D
326 Greenland	x	x					
328 Grenada	x	x	F		F	D	D
329 Guadeloupe	x	x					
258 Guatemala	x	x					
333 Guiana, French	x	x					
336 Guyana	x	x	F	D	F	D	D
263 Haiti	x	x					

Table 11.2

Trade Supplement Guide to the Coverage of Area and World Series

Country Code	Exports	Imports	Export Volume	Import Volume	Export Unit Value	Import Unit Value	Terms of Trade
614 Angola	x	x					
638 Benin	x	x					
618 Burundi	x	x	x	F	F	D	D
622 Cameroon	x	x					
624 Cape Verde	x	x					
626 Central African Rep.	x	x					
628 Chad	x	x					
632 Comoros	x	x					
634 Congo	x	x					
611 Djibouti	x	x					
642 Equatorial Guinea	x	x	x	D	x	x	D
644 Ethiopia	x	x					
646 Gabon	x	x					
648 Gambia, The	x	x					

Country Code	Exports	Imports	Export Volume	Import Volume	Export Unit Value	Import Unit Value	Terms of Trade
556 Maldives	x	x					
558 Nepal	x	x					
839 New Caledonia	x	x					
564 Pakistan	x	x	x	x	x	x	D
853 Papua New Guinea	x	x	x	D	D	x	D
566 Philipines	x	x	x	x	x	x	D
576 Singapore	x	x	x	x	x	x	D
813 Salomon Islands	x	x					
524 Sri Lanka	x	x	x	x	x	x	D
578 Thailand	x	x	x	x	x	x	D
866 Tonga	x	x					
846 Vanuatu	x	x					
582 Viet Nam	x	x					
862 Western Somoa	x	x	x	F	F	D	D

Country Code	Exports	Imports	Export Volume	Import Volume	Export Unit Value	Import Unit Value	Terms of Trade
268 Honduras	x	x	F	D	F	F	D
343 Jamaica	x	x					
349 Martinique	x	x					
273 Mexico	x	x					
353 Netherlands Antilles	x	x					
278 Nicaragua	x	x	F	F	D	F	D
283 Panama	x	x	F	F	F	F	D
288 Paraguay	x	x	F	D	F	F	D
293 Peru	x	x	F	D	F	D	D
362 St. Lucia	x	x					
364 St. Vincent	x	x					
366 Suriname	x	x	F	F	D	x	D
369 Trinidad and Tobago	x	x	x	x	x	x	D
298 Uruguay	x	x					

x Reported by country
F Calculated by IFS
D Calculated by Supplement

Table 11.3
Direction of Export Flows (% of Total)

		1966	1971	1976	1981	1981–1966
World						
Imports	BIL $	195.0	333.8	928.0	1,911.5	1,716.5
Industrial countries	%	73.7	75.3	68.5	65.7	− 8.0
Oil-exporting countries	%	6.6	7.3	15.0	15.2	8.6
Nonoil developing countries	%	19.0	16.2	16.2	17.3	− 1.7
Other countries	%	0.7	1.2	0.4	1.8	1.1
U.S.						
Imports	BIL $	27.8	48.8	132.5	273.3	245.5
Industrial countries	%	66.7	72.2	52.6	52.4	− 14.3
Oil-exporting countries	%	6.1	4.9	21.9	18.5	12.4
Nonoil developing countries	%	26.0	20.5	22.3	25.6	− 0.4
Other countries	%	1.2	2.3	3.2	3.5	2.3
Ratio of imports to GDP	%	3.7	4.6	7.8	9.5	5.8
Japan						
Imports	BIL $	1.1	19.8	64.9	142.9	141.8
Industrial countries	%	49.8	50.3	38.9	34.6	− 15.2
Oil-exporting countries	%	14.6	19.0	34.7	39.7	25.1
Nonoil developing countries	%	30.8	25.5	22.3	22.2	− 7.8
Other countries	%	5.6	5.2	4.0	3.5	− 2.1
Ratio of imports to GDP	%	9.0	8.6	11.6	12.6	3.6
Germany						
Imports	BIL $	18.2	34.5	88.4	163.9	145.7
Industrial countries	%	72.9	77.0	71.0	72.2	− 0.7
Oil-exporting countries	%	6.8	7.1	10.9	10.5	3.7
Nonoil developing countries	%	17.0	12.9	13.9	12.8	− 4.2
Other countries	%	3.3	3.1	4.2	4.5	1.2
Ratio of imports to GDP	%	14.9	15.9	19.8	23.8	8.9
Nonoil developing countries—Asia						
Imports	BIL $	14.8	21.7	57.7	164.7	149.9
Industrial countries	%	71.3	70.1	61.0	56.1	− 15.2
Oil-exporting countries	%	3.5	5.6	13.9	14.7	11.2
Nonoil developing countries	%	20.8	19.8	21.0	24.3	3.5
Other countries	%	4.5	4.5	4.2	4.8	0.3

Table 11.3 (*continued*)

		1966	1971	1976	1981	1981–1966
China PRC						
Imports	BIL $					
Industrial countries	%	75.8	74.6	79.7	72.9	− 2.9
Oil-exporting countries	%	1.5	0.9	0.4	1.0	− 0.5
Nonoil developing countries	%	22.7	24.5	19.8	26.2	3.5
Other countries	%	-	-	-	-	-
Hong Kong						
Imports	BIL $					
Industrial countries	%	52.7	61.2	52.2	50.4	− 2.3
Oil-exporting countries	%	2.4	2.2	1.2	0.6	− 1.8
Nonoil developing countries	%	42.4	30.7	38.8	40.6	− 1.8
Other countries	%	2.5	5.9	7.8	8.5	6.0
Ratio of Imports to GDP	%	90.0	92.6	91.7	102.3	12.3
Korea						
Imports	BIL $	0.7	2.4	8.8	26.1	25.4
Industrial countries	%	85.7	82.6	70.7	62.3	−23.4
Oil-exporting countries	%	2.7	7.9	20.8	23.5	20.8
Nonoil developing countries	%	10.0	7.7	7.2	9.0	− 1.0
Other countries	%	1.6	1.7	1.3	5.2	3.6
Ratio of imports to GDP	%	19.0	25.3	31.8	39.7	20.7
Singapore						
Imports	BIL $	1.3	2.8	9.1	27.6	26.3
Industrial countries	%	58.9	55.9	45.3	46.2	−12.7
Oil-exporting countries	%	13.3	9.3	22.4	25.4	12.1
Nonoil developing countries	%	26.5	32.1	29.3	25.9	− 0.6
Other countries	%	1.3	2.7	2.9	2.5	1.2
Ratio of imports to GDP	%	122.0	127.0	153.7	213.5	91.5

and the corresponding rise in the share of imports provided by the oil exporting countries have created substantial changes in the respective trade balances of these groups of countries (see table 11.4). In the 1950s, the industrial countries recorded deficits on their overall trade balances when the nonoil developing countries were turned into small and later sizable surpluses owing, in

Figure 11.1
Chart 2: Share of Industrial Countries' export receipts used to purchase imports
from the Developing Countries

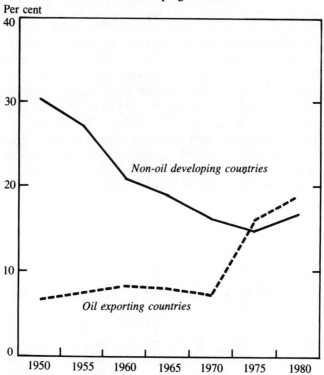

part, to the developing countries' continued efforts to industrialize through the *import of investment goods*. These trade surpluses, which went some way toward offsetting the large deficits with the oil exporting countries, also demonstrated the ability of the more advanced countries to expand export revenues faster than import payments. This reflected the fact that some of the traditional primary exports of the developing countries were being replaced by *similar goods produced by the industrial countries* themselves.[4] (emphasis added)

Figure 11.1 presents the share of industrial countries' export receipts used to purchase imports from developing countries, which have declined, as stated above, from 32.4 percent in 1950 to half its size, 16.7 percent in 1980.

4. *Supplement* op. cit., x, xi.

0.0 Minielement 4: Terms of Trade

The dependence of nonoil developing countries on their trade with the industrial countries is reflected in the trade balance figures in table 11.5. The industrial countries increased their positive trade balance with nonoil developing countries from $24.4 billion in 1979 to almost its double, $45.3 billion in 1981: which means that the nonoil developing countries developed a negative trade balance of the same magnitude with the industrial countries. The total negative trade balance of the nonoil developing countries in 1981 was US$ 109.0 billion.

Table 11.4 presents the annual percent change in the terms of trade. The terms of trade are closely linked to the price of primary commodities exported by the nonoil developing countries. In the second half of 1950s, the decrease of 3.4 percent in "all commodities" wholesale price index contributed to a 2.2 percent fall in the terms of trade of nonoil developing countries. During the '60s, the terms of trade rose marginally as the commodity prices increased. During the '70s, the commodity prices increased by more than 11 percent per annum, but it was offset by the rapid rise in industrial country export prices, resulting in a decline of 2.9 percent per annum in the first half of the 1970s. By 1980, the terms of trade were at their lowest in thirty years, as seen from figure 11.2.

0.0 Starting Negotiation Positions of CONCOL Players—Microelements

While the mini elements identify more specifically the perceived needs of the different countries with respect to international trade, the real *N*-positive indexes apply to the micro level, the level at which the specific high-technology and high-technology products enter into CONCOL negotiations.

0.0 Microelement 1: The U.S. Share of Country Exports

As the prime sources of high-technology imports, and as the prime target of high-tech products exports, the U.S. is viewed as the market that should be penetrated and held. The extent to which the U.S. accounts for the export earnings of a country, a nonoil developing country in particular is an *N*-positive index.

According to the Board of Trade of Taiwan, US$7.18 billion worth of exports, representing 44.4 percent of the total exports during January–August of 1983, went to the United States. Japan took 9.8 percent of the exports.

According to the Korean Office of Customs Administration, the U.S. accounted for US$6.24 billion worth of exports, representing 28.6 percent of the 1982 exports, with Japan buying 15.5 percent of the exports.

For Taiwan, 44.4 percent of the exports going to the U.S., and 9.8 percent to Japan could be considered the desired initial outcomes; for Korea, 28.6 percent of the exports going to the U.S., and 15.5 percent going to Japan could be considered the desired initial outcomes.

Table 11.4
Terms of Trade Developments
(Annual Per Cent Change)

	1950–55	1955–60	1960–65	1965–70	1970–75	1975–80	1976–81
Industrial Countries	0.1	1.4	0.5	0.4	−2.2	−2.1	−2.3
Oil Exporting Countries	N/A	N/A	−0.8	−1.4	27.3	11.4	13.7
Non-Oil Developing Countries	N/A	−2.2	0.1	1.1	−2.9	−0.9	−3.4
Africa	−2.1	−1.4	−0.1	2.4	−3.0	−3.5	−5.0
Asia	N/A	−0.7	−1.3	0.5	−4.7	−0.7	−3.0
Europe	N/A	N/A	N/A	N/A	−2.9	−1.5	N/A
Middle East	N/A	N/A	0.7	−0.8	−1.1	2.3	1.1
Western Hemisphere	N/A	N/A	N/A	1.1	−0.6	0.7	−5.4
Reference World trade commodity prices for major groups in U.S. dollars							
All Commodities	N/A	−3.4[1]	0.8	1.4	11.6	10.7	4.6
Food	N/A	−2.8[1]	1.1	2.4	17.8	6.5	7.7
Beverages	N/A	−7.9[1]	−1.3	4.3	7.4	17.5	−1.9
Agricultural Raw Materials	N/A	−1.2[1]	−2.4	−2.6	12.5	11.9	4.9
Metals	N/A	−2.7[1]	5.6	1.8	5.1	11.5	7.0
Petroleum[2]	2.5	−4.9	−2.4	−0.5	52.5	21.7	23.1

[1]1957–1960.
[2]Saudi Arabian light crude 34°–39° API gravity, average realized price, f.o.b. Ras Tanura. This series is not included in the "all commodities" idex.

Table 11.5
Distribution of Industrial Countries' Exports, Imports, and Balance of Trade*
(billions of U.S. dollars)

Year	Oil Exporting Countries			Non-Oil Developing Countries			World		
	Exports	Imports	Trade Balance	Exports	Imports	Trade Balance	Exports	Imports	Trade Balance
1950	1.6	2.4	-0.8	10.0	11.8	-1.8	36.4	39.8	-3.4
1955	2.8	4.0	-1.2	15.2	15.9	-0.7	58.5	61.8	-3.3
1960	4.4	6.8	-2.4	20.1	17.5	2.6	83.9	85.5	-1.6
1965	4.7	9.8	-5.1	27.1	23.7	3.4	125.7	131.4	-5.7
1970	7.6	15.7	-8.1	43.8	35.4	8.4	220.3	227.2	-6.9
1975	45.8	90.9	-45.1	113.2	83.9	29.3	569.0	589.5	-20.5
1979	75.0	166.1	-91.1	197.7	173.3	24.4	1,052.7	1,143.0	-90.3
1980	97.7	233.1	-135.4	240.6	206.9	33.7	1,239.4	1,369.6	-130.2
1981	115.2	214.0	-98.8	248.0	202.7	45.3	1,219.8	1,297.2	-77.4

Source: Direction of Trade Statistics data base. The world totals are the *IFS* totals shown elsewhere in the IMF *Supplement.*

Figure 11.2
Non-oil Developing Countries Trade Indexes
(1975 = 100)

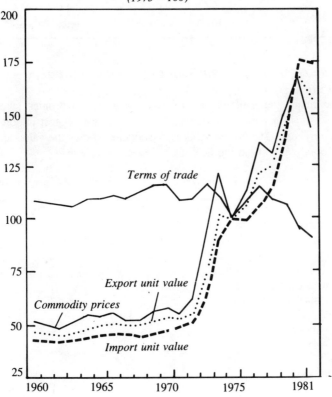

0.0 Microelement 2: The Profitability of Exports

Whether sold to the United States or to the rest of the world, exports should be profitable to the country concerned. A study published by the Korea Institute for Industrial Economics and Technology in May 1983 makes a comparative analysis of the export competitiveness of Korea, Japan, Taiwan, and Singapore. It derives a Profitability of Exports Index from the unit price of exports and production cost as follows:[5]

	1976	1977	1978	1979	1980	1981	1982	Average 1976-82
Korea	101.1	99.0	97.1	100.3	99.5	103.2	104.9	100.7
Japan	97.2	96.5	100.5	100.3	96.3	98.3	100.6	98.5

5. Unless otherwise specified, the comparative figures cited from the article in *Business Korea*, Vol. 1, No. 4 (October 1983), based on the Korea Institute Study.

Robotics/Artificial Intelligence/Productivity

	1976	1977	1978	1979	1980	1981	1982	Average 1976-82
Taiwan	98.6	96.9	99.1	99.0	92.8	89.0	94.5	95.7
Singapore	104.8	108.0	104.8	100.9	105.3	n.a.	n.a.	104.8

0.0 Microelement 3: Composition of Exports to the U.S. and to Japan

Among the major items of exports from Taiwan are: electronics, shoes, metal products, plastics, and plastic products. The U.S. is the biggest importer of Taiwan's leather shoes. The Korea Traders Association shows the value and composition of Korea's exports to the U.S. as follows:

	1979		1980		1981	
	US$ mil	%	YS$ mil	%	US$ mil	%
Total exports	4,379.1	100.0%	4,606.6	100.0%	5,560.9	100.0%
Agro-fishery products	143.0	3.3%	123.0	2.7%	133.9	2.4%
Raw materials and products of animal and plant origin	330.1	7.5%	270.6	5.9%	311.2	5.6%
Ore, fuels, and products of non-ferrous origin	44.4	1.0%	39.0	0.8%	61.2	1.1%
Chemicals and products	263.6	6.0%	265.6	5.8%	311.4	5.6%
Metals and steel products	742.3	17.0%	842.2	18.3%	1,094.7	19.7%
Machinery and carrier equipment	1,377.2	31.4%	1,149.4	25.0%	1,236.7	22.2%
Textiles	980.3	22.4%	1,113.5	24.2%	1,468.2	26.4%
Sundry goods	724.6	16.5%	804.3	17.5%	943.3	17.0%

The average share during the period 1979-81 is 26.2% for machinery and carrier equipment, 24.3% for textiles, 18.3% for metals and steel products, and 17.0% for sundry goods. These values can be considered the desired initial outcomes.

The 1979–81 averages for the major export commodities exported to Japan are 32.2% for textiles, 21.2% for agro-fishery products, 11.4% for machinery and carrier equipment, 11.3% for metals and steel products, and 7.3% for sundry goods.

With the exception of agro-fishery products, the leading export commodities to both the U.S. and Japan are identical.

During the first half of 1983, electronics rose to prominence. According to Korea Electronics Industries Cooperative (KEIC), the electronics industry's exports represent 11.5% of Korea's exports during January–June, 1983. Cathode ray tubes (CRTs) and semiconductors accounted for $382 million, radios, radio cassettes, and car radios accounted for $168 million, color TV receivers $141 million, hand-held calculators $30 million and electronic watches $24 million. The total of these major electronic exports of $793 million in the first half projects to an annual total of $1,586 million.

Constant comparison is made in Korea with Taiwan. In 1982, Taiwan reportedly exported US$2,984 million worth of electronics, compared with US$2,260 million exported by Korea.

0.0 Implications for the CONCOL Process of the Triad of CONCOL Elements

We have identified a triad of elements: macroelements, minielements, and microelements, in terms of which starting negotiating positions of CONCOL players can be specified.

In general, macroelements are N-zero; they are desirable goals, sanctioned by experience, but not unalterable. When we come to microelements, Korea would, for instance, find it essential to export 70.6 percent of its exports to industrial countries, and 14.6 percent to nonoil developing countries. Japan, on the other hand, may want to export double that percent to nonoil developing countries, 30.5 percent, cutting to two-thirds the export share of industrial countries, at 45.5 percent. As the IMF *Supplement* points out, the steady *decline* in the importance of the nonoil developing countries in the imports of the industrial countries is indeed a matter of serious concern to the nonoil developing countries. The importance of maintaining the share of the industrial countries of the exports from countries like Korea will make the N-Index positive, the export share being a *need*, much more than a *want*, making the need/want ratio positive.

The N-Index associated with the microelements is also positive, but the magnitude is probably much higher than the minielements. While the share of exports going to industrial countries is important, the share going to the U.S. is critical, making the N-Index higher in magnitude. Thus, the 28.6 percent of Korean exports going to the U.S. market is probably viewed more seriously by Korea than the 70.6 percent of Korean exports going to the industrial countries. Similarly, the average index of profitability of exports at 100.7 for the 1976–82 period is considered a standard that Korea should strive to keep and excel; and the 95.7 index is something that Taiwan would want to excel.

Of the 28.6 percent of exports from Korea going to the U.S., machinery and carrier equipment averaged more than a fourth at 26.2%, followed by textiles at

24.3% and metals and steel products at 18.3%. Each of these figures would be considered almost nonnegotiable in the sense of being inviolate.

0.0 Initial Desirable Outcomes from the U.S. Point of View

When Korea (and other developing countries, as well as industrial countries) seeks to expand exports of products, such as semiconductors—CRTs and semiconductors were the largest electronics exports in 1983—how would that jibe with the U.S.'s own industrial outlook?

0.0 U.S. Industrial Outlook of Ten Rapid Growth Industries

The U.S. Bureau of Industrial Economics studied forty-five manufacturing industries that grew rapidly between 1972 and 1978 and projected their future growth to 1985. The seven-year span contains a full business cycle. It includes the traumatic event of the 1973–74 energy crisis. In addition to these desirable features, the choice of the period 1972–78 was in part dictated by the fact that constant-dollar shipments were developed for the first time in 1972, the SIC system having been modified as of 1972. Table 11.6 presents the leading top growth industries.

0.0 Electronic Equipment and Components

The phrase "Electronic Equipment and Components" includes the following:

- 3662   Radio and televisions communication equipment
- 3671   Electron tubes
- 3674   Semiconductors and related devices
- 3675   Capacitors
- 3676   Resistors
- 3677   Coils and transformers
- 3678   Connectors
- 3679   Components not elsewhere classified

Of these, Semiconductors and related devices rank first among the rapid growth industries, with 19.1 percent compound average annual growth rate in 1972–78, and even higher 20.7 percent growth rate in 1979–85. The *U.S. Industrial Outlook* projects the shipments by the electronic equipment and component producers to resume the growth rate of the 1972–79 period during the years 1981–86. The compound annual rate of growth during 1972–81 of shipments was 9.0 percent for electronic equipment and components.

Exports of radio and television communications equipment and components amounted to $2.25 billion in 1981, which was an increase of 14 percent over 1980. A third of the exports goes to the United Kingdom, West Germany, Canada, and Japan.

Imports shot up by 41 percent over 1980 to $1.6 billion in 1981. As we noted

Table 11.6

Rapid Growth Industries and Growth Factors

SIC Code	Industry	Compound average Annual growth rates		New Products	Less than avg. price increase	Product Substitutions	Expanded Export Market	Energy-related	Other
		1972–78	1979–85						
3674	Semiconductors and related devices	19.1	20.7	X	X	X	X		
3693	X-ray apparatus and tubes	19.1	6.5	X			X		
3832	Optical instruments and lenses	18.4	13.8	X	X		X		
3573	Electronic computing equipment	17.0	15.0	X	X		X		
3795	Tanks and tank components	15.7	15.7						X
3574	Calculating and accounting machinery	13.6	10.0	X	X	X	X		
3579	Office machines not elsewhere classified (n.e.c.)	9.7	15.0	X		X	X		X
3823	Process control instruments	11.0	12.9	X	X	X	X	X	
3822	Environmental controls	6.5	11.9	X				X	
3829	Measuring and controlling devices, n.e.c.	8.4	10.6	X				X	

Source: 1982 U.S. Industrial Outlook, (Washington, D.C.: U.S. Department Commerce, January 1982).

earlier, Korean exports of radio, radio cassettes, and car radios amounted to $168 million in the first half of 1983, while CRTs and semiconductors accounted for $382 million. The U.S. is, of course, the prime target market.

Turning from radios to components, $6.2 billion worth of exports is projected for 1982. "Over 70 percent of the value of electronic component exports consists of semiconductor parts sent to offshore assembly facilities. A large number of the finished semiconductors come back to the United States for sale in the domestic market."[6]

> Imports in 1982 are projected at $5.8 billion.
>
> Imports of components are dominated by semiconductors which account for 67 percent of the total. After assembly abroad, most of these devices enter the United States under the tariff provision of items 806.30 and 807.00. The value of these 806/807 imports was estimated at $2.5 billion in 1980 and 1981. Under these tariff provisions only the value added abroad portion of imports is subject to duty. . . .
>
> *Semiconductors will continue as the dominant factor in equipment technology developments at least through the 1980s.* Such devices as very high speed integrated circuits (VHSIC) are now being developed under a Defense Department program.[7] (emphasis added)

0.0 Differentiation in Semiconductors

Integrated circuits are clearly the dominant element of semiconductors exports and imports. The race to fabricate Very Large Scale Integrated Circuits (VLSI) is on; not only in the United States and Japan, but in European countries as well.

Integrated circuits trade with Japan shows that there was a modest surplus before 1978, but during 1978–81, it turned to a deficit three times as large at $151.9 million. The important fact is that the imports of ICs reflect the sophistication of the needs they fulfill. As the sophistication increases, so will the differential advantages. It is conceivable that Country A may import from Country B certain types of VLSIs, imbed them in a product and export the product to Country B (or to Country C, D, E) which also manufactures VLSIs. If the degree of sophistication of VLSIs varies, it is possible that Country A may import VLSI-I from Country B, VLSI-II from Country C, etc., and export products to Country B, C, etc.

6. *U. S. Industrial Outlook 1982*, (Washington, D.C.: Department of Commerce, 1982), 232.
7. *Ibid* 233.

0.0 Strategy of Selecting the Industrial Structure

The race to VLSI signals a fundamental change in comparative advantage. As Robert Reich puts it:

> Just as the main source of comparative advantage changed over a century ago from static natural endowments to slowly accumulated capital stocks, so now the new importance of skill-intensive production makes comparative advantage a matter of developing and deploying human capital. This second change is more dramatic than the first. In a very real and immediate way, a nation *chooses* its comparative advantage. . . .
>
> Most discussions of Japan's competitive success focus, either admiringly or accusingly, on its tactics, while neglecting the fact that these tactics are effective largely because they are rooted in a coherent strategy for progressively adopting higher-skilled, higher-valued economic activities.
>
> As Japan reduced its commitment to basic steel, basic petrochemicals, small appliances, ships, and simple fibers, it has dramatically expanded its capacity in the higher-valued, more specialized segments of these industries.[8]

Reich argues that skilled labor has become the only dimension of production in which advanced industrialized nations can create and retain an advantage—precision castings, specialty steel, special chemicals, sensor devices, design and manufacture of fiber-optic cable, fine ceramics, lasers, large-scale integrated circuits and advanced aircraft engines. In this context, it is no longer meaningful to talk monolithically about industries such as "steel," "textiles," or "semiconductors." Segmentation of each industry into activities that provide higher-value-added segments *and more competitive outputs become the key to a successful CONCOL strategy* in which, for instance, Country A cooperates with Country B in certain segments, while competing against in the other. We explore this in greater detail with respect to VLSI and Fifth-Generation Computers in the next chapter.

8. Robert B. Reich, "Beyond Free Trade," *Foreign Affairs* (June 1983), 782.

12

Cars and Computers: Keys to a U.S.-Japan CONCOL

OVERVIEW: Given the *initial* desirable outcome in international trade in terms of seven macroelements, four minielements, and three microelements, which provide the context of the CONCOL process, what are the specific elements that enter into U.S.-Japan and other CONOLs? Crucial to all U.S.-Japan CONCOLs is the Fifth-Generation (5G) Computer technology, using the term to symbolize the decisive change in technology brought about by design in Japan.

The 5G is not a computer; it is a revolution. To appreciate the revolutionary potentials of 5G, computer technology development is briefly reviewed in terms of three phases: from data processing to information processing; from information processing to decision support systems; and from decision support systems to knowledge-processing systems. The 5G will be a knowledge-processing machine by virtue of three interrelated capabilities: understanding speech, graphics, and natural language; choosing the most appropriate knowledge-base; and solving problems. While progress is being made toward realizing these capabilities, attention must be focussed on 5G-based products.

The revolutionary potentials of 5G technology are three-fold: manufacturing, marketing, and management. In manufacturing, 5G holds promise in three areas —improved quality, innovative quality, and individualized quality. These capabilities will impact the marketing of 5G technology products. The impact could be specified in terms of the ratio of performance specific (PS) products to performance general (PG) products, the latter being items of mass manufacture, and the former, custom-manufactured items. As affluence increases, demand tends to shift from PG to PS products, both in the domestic market and in foreign markets.

When and where should PG and PS products be made and sold? Five critical factors in the optimum selection of the product mix and manufacturing location are: (1) sales volume, (2) profit volume, in units, (3) profit units as percent of total, (4) PS/PG ratio of units sold, and (5) PS/PG ratio of profit. These are applied to a detailed illustration of the manufacture and marketing of cars under 5G technology in five markets: (1) domestic, (2) industrial country 1, (3) industrial country 2, (4) nonoil developing country 1, and (5) nonoil developing country 2. The

third area of revolutionary potential from 5G technology is management. While the least visible at the moment, these potentials are likely to exert a profound effect on how the world thinks and trades, plays and lives in the coming decades.

0.0 Japan's Fifth Generation (5G) Computer As a Centerpiece of U.S.-Japan CONCOL

The 5G is not a computer; it is a revolution.

After two years of preliminary investigations, the Japanese Government launched, in April 1982, the 5G Project. Its aim is to develop by 1990 a prototype 5G computer which will be a knowledge-processing system. Such a computer will revolutionize the products and services of the last decade of the twentieth century probably more dramatically than the Industrial Revolution did in the latter half of the eighteenth century. The incredible internationalization of production and sales, which make almost commonplace the selective assemblage, in different corners of the earth, of talent and technology to fabricate and devise products and services to be sold in other corners of the earth, thrusts to the forefront the paramount issue of territory, in the sense of commercial territory—the market for the high-technology products. It is as though the market comes first; and the products second—or, territory first, technology, second. Bruising battles for the markets are a genuine possibility when the U.S. and Japan (and other trading partners) each push further, farther and faster the high-technology frontier on its own, and while the technology is yet a gleam in the eye of the inventor and the innovator, vie with one another for global markets for high-technology products that have yet to see the light of day.

The thesis of this chapter is that a U.S.-Japan CONCOL on 5G-type technology is critical to the healthy progress in trade and technology of both countries.

0.0 From Data Processing to Information Processing

To appreciate the truly revolutionary impact of the 5G (which we will use to denote generically exceptionally significant developments in high-technology), let us look briefly at the evolution of computer technology developments.

The Electronic Numerical Integrator and Calculator (ENIAC) ushered in the era of data processing. What filled a large room at the University of Pennsylvania in 1946 did much less and at much greater cost than what something that occupies the space of half a thumb does today, the computer-on-a-chip being much faster and much more economical than the computer-in-a-room.

In data processing, data are the input; and *data* are also the output.

The promise of computers, however, is *not* in replacing the ancient abacus by a modern one, in the sense of doing faster the very same thing that slower processors of data do. Can something more be derived from the data than the same inputs in another form?

That is the question of information processing. In information processing, data are the input; but *information* is the output.

Information is data that affect behavior. To affect behavior, data have to be processed in relation to one's goals. Alternative actions would result in advancing toward one's goals or in moving away from them. Given such alternatives, one can change the outcome by changing the action. Information-processing systems are in great demand as decisions become more complex.

0.0 From Information-Processing to Decision Support Systems

Much has been promised in the name of information-processing systems: Enough promises to make the phrase "Management Information Systems" one either to swear by or to swear at. While the three words that make up the phrase have themselves been in use for a much longer time, Management Information Systems (MIS) as currently understood would go back only about a decade.

It is clear, however, that the promise of MIS has not been universally realized. In fact, there have even been suggestions that more often than not, MISinformation is what has been generated.

One of the reasons why MIS has been popular is because those who have to consider a large number of interacting variables, each taking a range of values, could not cope with them on the back of an envelope. They could visualize a giant computer spewing out in a matter of minutes measures of relationships that ordinary mortals could not even write out in a lifetime; and given alternate rules for ordering them, could spew out other sets of numbers just as easily.

What went amiss is that those who could coax out of the computer the myriads of values and relationships did not often command the convictions of the decision-makers. The latter did not participate in the models that the computer whizzes whipped out, and were naturally wary of the electronic oracle.

Even the minimum knowledge requirements of computerspeak have consistently tended to make the executive computer-shy. No less a citadel of computers than the headquarters of IBM found that placing a terminal in every manager's office far from assured their prompt and faithful use. On the contrary, faced with positive nonuse of the terminals, IBM withdrew them from all the managers' offices in the early seventies.

A technological development that originated in 1971, but that is only now making its widespread accessibility credible, is the microcomputer. The computer-on-a-chip suggests that the computer is more readily accessible to an individual, encouraging an executive to carry out his trial and error in privacy. Further, the development of high-order languages has made it possible for the computer-shy executive to use nearly humanspeak in commanding the electronic slave to do his or her bidding.

Recognizing that the gallop of computer printouts in the name of MIS was somewhat like that of a riderless horse, a chorus has been rising to the theme: put the decision-maker in the saddle. Instead of astounding him with answers, let

him ask the questions. If he does not frame them in machinespeak, then develop humanspeak, which he can use, and have it translated into machinespeak. Even more daring, let the decision-maker suggest what the reality is like, so that he or she will in fact be building a model of reality. If the manipulation of the model does not look right the first time, let the decision-maker suggest what variables should be changed, and how, and try again, and again, until a model that he or she can use should be developed. The key is: *adaptive design and usage*.

0.0 Types of Decisions—Context E, Context T

What types of problems should be subject to such involvement by the decision-maker?

The routine and the repetitive decisions need not involve the high-level decision-makers. The very routine nature of the decisions suggests that not only the variables, but also their interactions, are well understood to develop decision rules which can be repeatedly applied.

The decisions are primarily with respect to the external *demand* for the products and services of the organization, or with respect to the internal *supply* of capabilities to meet the demand. We refer to the former as Context E-Environmental Status of Fulfillment Characteristics, and the latter as Context T-Technological Status of Performance Characteristics.[1]

Context E is characterized in terms of the distribution of the demand for the products and services of the organism—a term that underscores the interaction of human beings in an organization giving rise *collectively* to an output more than the sum of their individual capabilities, or to less than the sum of their individual capabilities, but rarely equal to the sum. The frequency of the demand for specified numbers of units, e.g., 0-1,000, 1,0001-10,000, 10,0001-100,000, etc., is known from past data. Given the frequency of these distributions, a smoothed curve can be fitted to the data, formally known as the *probability distribution*. Its name stems from the fact that the probability of the demand being equal to, less than, or greater than a given quantity, say X, can be determined from the fitted curve. To fit the curve to the data, we need to know the form of the distribution, and we need to know the parameters of the distribution, e.g., mean and standard deviation, if the form is the familiar bell curve.

In the case of products and services with established history, e.g., the demand for light bulbs, we know quite well the form of the distribution; and we can determine the parameters from the data. In other words, both the form and parameters of the fulfillment characteristics are known. We use the term "fulfilment" because the products and services are fulfilling the perceived wants and needs of the customers. At the other extreme is the case of a product so new that even the form of the demand distribution is unknown, making the parameters, of course, also

1. George K. Chacko, *Management Information Systems* (Princeton, N.J.: Petrocelli Books, 1979), ch. 3.

unknown. In between, we have demand distributions whose form is known (or could be guessed at intelligently), but whose parameters are unknown. The three Context E situations are named as follows:

Context E	Form of the Distribution	Parameters of the Distribution
Uncertain	unknown	unknown
Risk	known	unknown
Reduced risk	known	known

The decisions made by the decision-maker are with reference to both Context E and Context T. Context T characterizes the *internal* capabilities of the organism to meet the external demand. If the technology is unstable, as is most often the case with new developments, e.g., the semiconductor technology in the early sixties, the products and services that the organism could offer are subject to the unstable technology. If one cannot promise a particular product, it would be pretty hard for the customer to commit himself to buying the unknown product, often making the unstable Context T be associated with uncertain Context E. At the other extreme, Context T is quite stable because the product has been made the same way year in and year out for many years. Associated with the stable Context T is usually reduced risk Context E. In between, we have quasi-stable technology which can be associated with risk Context E.

We can give *operational* content to Context T by using the concept of statistical control. When the production process is in statistical control, the technology is stable. When the production process is in statistical control, each item can be expected to meet the standards so well that only a small fraction, e.g., a hundred light bulbs out of a hundred thousand, needs to be inspected to assure one of the quality of the whole lot. In addition to the inspection by the producer, the customer may also inspect the lot. The total number of inspections performed by the producer and the consumer together on the lot of one hundred thousand light bulbs, say two hundred, is a very small fraction of the total, such as one or two in a 1,000. The ratio of one in a thousand can be expressed as 10^{-3}, the exponent being *negative* in this case, or *M*-Negative.

At the other extreme, when the technology is unstable, the producer himself will inspect each item perhaps several times to make sure that the product does indeed meet specifications. Further, warned that the technology is unstable, the customer is also bound to inspect each product one or more times. The total number of inspections—by the producer and the consumer together—is larger than the number of products itself, making the ratio of number of inspections to the number of units greater than one. For instance, if the number of inspections of each product is one hundred, the ratio is 10^2, the exponent in 10^M being *positive* in this case, or *M*-Positive.

In between the extreme cases in which the exponent is negative or positive is

the instance in which the exponent is zero. Instead of inspecting each item several times, or only fractionally, it can be inspected just once, making the exponent in 10^M *zero* in 10^0, or *M*-Zero.

We can classify the decision-making systems as follows:

	Unstable	Quasi-stable	Stable
Uncertain	*M*-Positive		
Risk		*M*-Zero	
Reduced risk			*M*-Negative

What types of decision situations merit Decision Support Systems?

The discussion on the subject uses several terms, but the intent can be summarized as follows:

A Decision Support System (DSS) ought to support decisions of the *M*-Positive type; and ought *not* to bother with the decisions of the *M*-Negative type. Some of the *M*-Zero type decisions merit DSS; some others do not.

While the usual associations of Context E and Context T are as shown above, other combinations are feasible:

Context T ⟍ Context E	Unstable	Quasi-stable	Stable
Uncertain	*M*-Positive	*M*-Zero	*M*-Negative
Risk	*M*-Positive	*M*-Zero	*M*-Negative
Reduce risk	*M*-Positive	*M*-Zero	*M*-Negative

We offer the following definition of DSS:

> *The Decision Support Systems (DSS) is basically a computer-based aid, including algorithms and interfaces, primarily to assist the user in committing resources to achieve results in situations with little prior applicable experience, by his or her progressively discovering alternative resource-results combinations corresponding to changing priorities of objectives.*

0.0 From Decision Support Systems to Knowledge-Processing Systems

We mentioned that in data processing systems, data are the input and data are also the output. In knowledge-processing systems, data are the input, but knowledge is the output.

What is knowledge?

In the personal discussions with Dr. Tohru Moto-oka, chairman of the Committee for Study and Research on Fifth-Generation Computers, the recurring aspect of knowledge was that of inference—given many assertions and rules, coming up with *all* the valid conclusions, and *only* the valid conclusions.

Our reluctant executive need no longer be computer-shy; he or she may never have to type a word or punch a card. The executive can simply *say* the word in Japanese, in English, or in another *natural* language; and the 5G listens and obeys. If the executive would rather draw, that is okay with 5G; it will decipher the master's hieroglyphics.

The 5G will be a knowledge-processing machine because it can do three interrelated things. One, it can understand speech, graphics, and natural languages (intelligent interface). Two, it can convert the incomplete description of the human user into a complete description, choosing from the many stores of varieties of knowledge the most appropriate (knowledge-base management). Three, it can solve the problem and draw the right conclusions (problem-solving and inference functions).

We speak of these three interrelated functions as though they are already accomplished facts.

They are not accomplished facts, but targeted objectives.

For purposes of CONCOL discussions, however, we need to treat them as accomplishable facts and raise the question: What type of products are likely to flow out of 5G? How could the pursuit of 5G technology itself be shared between the U.S. and Japan during an immediate period, such as 1985–1990, so that they can preagree on the 5G products and services trade during a subsequent period, say, 1991–2000?

0.0 Revolutionary Potentials of 5G Technology: Manufacturing

The main areas of the revolutionary potential of 5G technology are threefold— manufacturing, marketing, and management. In the context of a declining rise in U.S. productivity and of rising Japanese productivity, it is but natural that attention will first be turned to manufacturing. The technology of the Fifth Generation holds promise in three areas: improved quality, innovative quality, and individualized quality.

Robotics/Artificial Intelligence/Productivity

0.0 Improved Quality in Performance General (PG) Products—Today's Cars

At the current state-of-the-art of robotics, precision performance by a million cars off the robotics factory is quite feasible. Every one of the million cars will conform within very small tolerances to characteristics which are general to all of them. The gear-ratio, the drag coefficient, the acceleration, the deceleration, etc. of car number 9 will be hardly distinguishable from car number 1,457 or from car number 99,999, or any other of the cars from one to a million. All of them exhibit general performance characteristics in the sense that the performance characteristics are common to all the million cars, and not specific to any one of them—or, the million cars are *performance general products*.

Testimony to the significant success of Japan in providing performance general products is reflected in the fact that the imports of Japanese cars into the U.S. limited by hard bilateral negotiations to 1.68 million units under a three-year agreement ending in March 1984, had to be raised to 1.8 million for 1985. In other words, the auto PG products that the United States offers are so much less acceptable to the U.S. consumers than what Japan offers that their imports have to be restricted. William G. Ouchi, author of Theory Z, talks about a team of engineers and managers from Buick visiting their dealer in Tokyo. The American visitors, seeing a massive repair facility, asked how the Tokyo dealer built up such a large service business. The dealer replied that it was *not* a repair facility at all. With some embarrassment he explained that it was a *reassembly operation* where the newly delivered American cars had to be disassembled and rebuilt to meet Japanese standards.

Ouchi makes the point that there is a happy ending to the story about the poor quality of American Buicks. By involving the workers in the production process—a key ingredient of Theory Z—and redesigning the management of that plant in ways that resemble the Japanese approach, the Buick plant was able to rise to the number one position among all General Motors plants in quality and efficiency in two years.

0.0 Innovative Quality in Performance General (PG) Products—Tomorrow's Cars

Tomorrow's cars will depend heavily on installed computers. They will control engine performance and chassis suspension, monitor fuel consumption, warn the driver of hazards, and inform the driver of the location of the car on the highway, etc. The computer-aided car will benefit greatly from the progress in 5G technology—intelligent interface, knowledge-base management, and inference.

Again, these innovative features will be identical in all the units that come off the assembly line; hence we treat them as PG.

300

0.0 Individualized Quality in Performance Specific (PS) Products—Customized Cars

Even with PG products, a certain amount of customization takes place. Oldsmobile coaxes prospective customers by saying: "Can we build one for you?" In this case, the customization is on the color of the car, color of the upholstery, color of the tires, etc.

What 5G technology will make possible is the customization of car performance characteristics themselves. Now the province of the specialist and the enthusiast, car engine performance can indeed become the province of virtually anyone who wants to trade comfort for efficiency, or even raise both at some cost. Already, the advanced computer-aided models can determine the gear-ratios for different segments of the trip. It is not inconceivable that in the near future, trip-routing and speed selection could be done by the car computer system. The customization *at the time of purchase* could be governed by the various trade-offs that the customer is willing to make, as well as the provision he wants to make for further customization at a later date. The more the customization, the less the car will share characteristics with other cars; they are, therefore, *performance specific products*.

0.0 Revolutionary Potentials of 5G Technology: Marketing

The proof of the pudding is in the eating; the proof of the technology is in the trade.

The technology of the Fifth Generation promises to impact marketing, both national and international, in a number of areas: (1) PSPG ratio of present individual products—national market, (2) PSPG ratio of present individual products—international markets, (3) PSPG ratio of present group(s) of products—national, (4) PSPG ratio of present group(s) of products—international, (5) PSPG ratio of potential products—national, and (6) PSPG ratio of potential products—international.

0.0 PSPG Ratio of Present Individual Products—National Market

The technical capability of customizing performance characteristics that 5G technology offers will necessitate segmenting the market for the product. For instance, it is quite conceivable that on the basis of the specific driving needs of the person, as shown by the record of driving over a period of time, 5G technology can "build one for you"—design a new car altogether, while optimizing fuel efficiency and comfort in the present car.

The new car thus designed will be performance-specific (PS).

The *national* market for cars will consist of those who order PS and those who order PG cars. If one hundred cars are produced, le us say that 68 percent is

sold domestically, and 32 percent exported. The PS/PG ratio of cars in the domestic market may not be the same as abroad. Further, the foreign market is itself differentiated—industrial countries, oil-exporting countries, nonoil developing countries, and other countries will each have its own PS/PG ratio, not to speak of the PS/PG ratios of individual countries within each group.

0.0 PSPG Ratio of Present Individual Products—International Markets

With respect to customized cars, it is probable that the PS/PG ratio will be higher in industrial countries than in nonoil developing countries. Owing to the wide disparity in income in the developing countries, PS cars may well be demanded by the very rich in these countries while PG cars are demanded by the less wealthy. The different (PS/PG) ratios could be as follows:

Market	PS Units	PG Units	(PS/PG) × 100
Domestic	18	50	36.0
Industrial countries:			
Country 1 (IC1)	2	7	28.6
Country 2 (IC2)	2	5	40.0
Nonoil developing countries			
Country 1 (DC1)	2	8	25.0
Country 2 (DC2)	1	5	20.0
Total	25	75	33.3

0.0 PSPG Ratios and Profit Share of Individual Products—International Markets

It is likely that the PS products will be more expensive and more profitable than the PG products. For instance, PS could be 50 percent more profitable than the PG products. Let us say that the profit from each PG product is 1 unit, and that from each PS product is 1.5 units. The total profit from PG products is (75 × 1 =) 75 units, while the total profit from PS products is (25 × 1.5 =) 37.5 units.

The profit share of PS products is (37.5/(75 + 37.5) =) 33.3% and that of PG products is (75/(75 + 37.5 =) 66.7%. The profit shares of each market segment identified earlier are as follows:

Market	$\frac{\text{PS Profit}}{\text{Total}}$	$\frac{\text{PG Profit}}{\text{Total}}$	$\frac{\text{PS Profit}}{\text{PG Profit}}100$
Domestic	24.0%	44.4%	54.0%
Industrial countries:			
Country 1 (IC1)	2.7%	6.2%	43.5%

Country 2 (IC2)	2.7%	4.4%	61.4%
Nonoil developing countries			
Country 1 (DC1)	2.7%	7.2%	38.0%
Country 2 (DC2)	1.3%	4.4%	29.5%
	33.3%	66.7%	50.0%

0.0 Stages of Growth of Individual Products—National and International Markets

Each product that is born, dies. Before dying, it declines. It begins to decline after it matures. Before maturing, it goes through rapid growth. Prior to rapid growth is infancy, which follows birth, coming after a period of gestation.

Unless catastrophe intervenes, each product goes through the life cycle stages of gestation, birth, infancy, rapid growth, maturation, decline, and death.

With respect to each product with PS and PG components that are marketed both domestically and internationally, the stages of the life cycle experienced must be calibrated. What is more, the stages of growth should be *anticipated*. Thus, it is essential to know when the PS component is about to enter rapid growth/maturity/decline; similarly, the PG component. The knowledge that rapid growth is about to be entered signals the need to step up production in either PS or PG production, or both. Similarly, the imminent entry into maturation signals the need to cut back production.

In the illustrative example, there are twelve (12) different life cycles: (1) domestic PS market, (2) domestic PG market, (3) IC1 PS market, (4) IC1 PG market, (5) IC2 PS market, (6) IC2 PG market, (7) DC1 PS market, (8) DC1 PG market, (9) DC2 PS market, (10) DG2 PG market, (11) Total PS market, and (12) Total PG market.

Looking at the total market, we may say that the idea of customizing cars with reference to performance characteristics (and not merely the externals and accessories) will be sufficiently unfamiliar to make the demand for PS cars relatively smaller than the demand for PG cars at the outset. In other words, the PS/PG ratio will be quite small in the infancy stage of the total market for the product. As the idea of customization catches on, more and more people would opt for the PS car, making the ratio PS/PG ratio larger. When would the manufacturer stop making that type of PS offering? Probably when the number of people buying PS cars is equal to the number of people buying PG cars, or the PS/PG ratio is 1. Why? At that point, it would be economical to mass-produce the PS options and offer at even lower price the set of customizations which make the PS distinct from the PG version of the same product, the car in this instance. The time would be ripe then to offer a different set of customizations that would make the new PS version different enough from the old PS version which is now the PG version. The cycle

starts again with the *new* PS and the new PG (which is the old PS, mass-produced).

A plausible profile of the life cycle can be sketched with reference to the PS/PG ratio of the total market.

Stage	Market	PS Units	PG Units	(PS/PG ratio) × 100
A	Total market	25	75	33.3
B	Total market	40	60	66.7
C	Total market	50	50	100.0

0.0 Strategy of International Trade Shifts in Technology and Trade—Individual Products

We have designated the Stages as A, B, and C because we are in effect making Stage C the maturation stage. Instead of letting the product enter the stage of decline, what we are doing is making the maturation stage of the first PS product become the infancy stage of the second PS product, by redefining what constitutes the customization, and thereby introducing a new product, PS1. The "1" denotes the changes in customization that will constitute a sufficiently different set of performance characteristics from PS to merit being considered a new product.

In table 12.1 we illustrate Stages A, B, and C of PS and PG versions of the car in the three principal world market areas: domestic, industrial countries, and nonoil developing countries.

As important as the units are, the profits—both absolute and relative—we, suggest that the unit profit of PS could be 150 percent of PG. If we further consider that Stage A is the initial stage of the PS and PG versions of the car, the profit picture would be expected to change. As PS becomes more popular, and the manufacturer gains more experience, the unit cost would be expected to drop perhaps to 120 percent of PG in Stage B. and to 110 percent in Stage C. A further decline to 105 percent in Stage D is consistent with this pattern.

In table 12.2 we present the illustrative figures of unit sales, profit units, and profit share of both PS and PG by the various markets.

How should the manufacturer proceed with the different markets for PS and PG versions of the car?

Five factors are critical to the optimum selection: (1) sales volume in units, (2) profit volume in units, (3) profit units as percent of the total, (4) PS/PG ratio of units sold, and (5) PS/PG ratio of profit percent.

In table 12.3 we present the values of the four factors by product and market. Earlier we suggested that when PS/PG ratio reaches 1.00, or the (PS/PG) × 100 reaches 100, it is time to introduce the new product, PS1, so that what would otherwise have been the decline stage for PS would now be converted to the infancy stage for the new product PS1.

Table 12.1

Illustrative (PS/PG) Ratios of Cars by Stages of Growth

Market	PS Units	PG Units	(PS/PG) × 100
Stage A			
Domestic	18	50	36.0
Industrial countries:			
Country 1 (IC1)	2	7	28.6
Country 2 (IC2)	2	5	40.0
Nonoil developing countries:			
Country 1 (DC1)	2	8	25.0
Country 2 (DC2)	1	5	20.0
Total	25	75	33.3
Stage B			
Domestic	14	20	70.0
Industrial countries:			
Country 1 (IC1)	10	15	66.7
Country 2 (IC2)	8	11	72.7
Nonoil developing countries:			
Country 1 (DC1)	5	8	62.5
Country 2 (DC2)	3	6	50.0
Total	40	60	66.7
Stage C			
Domestic	12	5	240.0
Industrial countries:			
Country 1 (IC1)	18	22	81.8
Country 2 (IC2)	9	5	180.0
Nonoil developing countries:			
Country 1 (DC1)	7	8	87.5
Country 2 (DC2)	4	10	40.0
Total	50	50	100.0

The ((PS/PG) × 100) is of course one of the four factors that should enter into the development of a marketing strategy. It is quite conceivable that we could start PS1 in say, the domestic market, while continuing to expand the PS market in say, Industrial Country 1, Developing Country 1, and Developing Country 2.

Table 12.2
Illsutrative Sales and Profits of Cars by Markets by Stage of Growth

Market	CarType	Stage A			Stage B			Stage C		
		Volume	Profit Units	Profit %	Volume	Profit Units	Profit %	Volume	Profit Units	Profit %
	PS	18	27	24.0%	14	16.8	15.5%	12	13.2	12.6%
Domestic	PG	50	50	44.4%	20	20.0	18.5%	5	5.0	4.8%
		68	77	68.4%	34	36.8	34.0%	17	18.2	17.4%
Industrial Countries:										
IC1	PS	2	3	2.7%	10	12.0	11.1%	18	19.8	18.9%
	PG	7	7	6.2%	15	15.0	13.9%	22	22.0	20.9
		9	10	8.9%	25	27.0	25.0%	40	41.8	39.8%
IC2	PS	2	3	2.7%	8	9.6	8.9%	9	9.9	9.4%
	PG	5	5	4.4%	11	11.0	10.2%	5	5.0	4.8%
		7	8	7.1%	19	20.6	19.1%	14	14.9	14.2%

Nonoil developing countries:

DC1	PS	2	3	2.7%	5	6.0	5.6%	7	7.7	7.3%
	PG	8	8	7.2%	8	8.0	7.4%	8	8.0	7.6%
		10	11	9.9%	13	14.0	13.0%	15	15.7	14.9%
DC2	PS	1	1.5	1.3%	3	3.6	3.3%	4	4.4	4.2%
	PG	5	5.0	4.4%	6	6.0	5.6%	10	10.0	9.5%
		6	6.5	5.7%	9	9.6	8.9%	14	14.4	13.7%
Total	PS	25	37.5	33.3%	40	48.0	44.4%	50	55.0	52.4%
	PG	75	75.0	66.7%	60	60.0	55.6%	50	50.0	47.6%
		100	112.5	100.0%	100	108.0	100.0%	100	105.0	100.0%

Table 12.3
Illustrative Sales, Profit, and (PS/PG) Ratios by Stage of Growth

Market	Growth Stage	Volume	Profit Units	Profit %	(PS/PG) × 100 Units	(PS/PG) × 100 Profit %	
Domestic	A	68%	77	68.4%	36	54.0	
	B	34%	36.8	34.0%	70	83.8	
	C	17%	18.2	17.4%	240	262.5	Start PS 1 Manufacture
IC1	A	9%	10	8.9%	28.6	43.5	
	B	25%	27	25.0%	66.7	79.9	
	C	40%	41.8	39.8%	81.8	90.4	Start PG manufacute
IC2	A	7%	8	7.1%	40.0	61.4	
	B	19%	20.6	19.1%	72.7	87.2	
	C	14%	14.9	14.2%	180.0	195.8	Start PS Manufacture
DC1	A	10%	11	9.9%	25.0	42.7	
	B	13%	14	13.0%	62.5	75.7	
	C	15%	15.7	14.9%	87.5	96.0	
DC2	A	6%	6.5	5.7%	20.0	29.5	
	B	9%	9.6	8.9%	50.0	58.9	
	C	14%	14.4	13.7%	40.0	44.2	
Total	A	100%	112.5	100.0%	33.3	49.9	
	B	100%	108.0	100.0%	66.7	79.9	
	C	100%	105.0	100.0%	100.0	110.1	

Insofar as PS1 represents more advanced technology than PS, the less advanced PS technology may be transferred to other countries under licensing or joint venture agreements. The manufacturing and the marketing of PS, PG, and PS1 could be as follows:

Market Segment	Stage A Manufacture	Stage A Market	Stage B Manufacture	Stage B Market	Stage C Manufacture	Stage C Market
PS1						
Domestic	PG	PG	PG	PG		PG
	PS	PS	PS	PS	PS	PS
IC1		PG		PG	PG	PG
		PS		PS		PS
IC2		PG		PG		PG
		PC		PS	PS	PS
					PS1	PS1
DC1		PG		PG		PG
		PS		PS		PS
DC2		PG		PG		PG
		PS		PS		PS

0.00 Shift to New Technology—Manufacturing for Domestic Market

We have only three data points, Stage A, Stage B, and Stage C. What do they indicate in table 12.3 about the *domestic* market as far as the need for new technology is concerned?

1. The volume has decreased from 68 percent to 17 percent.
2. The profit has decreased from 77 to 18.2 units.
3. The share of the profit has decreased from 68.4 percent to 17.4 percent.
4. The ratio index of (PS/PG) units has risen from 36 to 240.
5. The ratio index of (PS/PG) profit % has risen from 54 to 262.5.

While the domestic market is only a quarter as important in Stage C as it was in Stage A, the importance of PS to the domestic market has risen dramatically. To cater to the domestic market, which finds the PS-type technology meeting its needs, the domestic market should introduce the next level of PS technology, viz., PS1. This is indicated by listing PS1 against "Domestic Market" under "Manufacture" and "Market" in the table above.

0.0 Shift to New Technology, PS—Manufacturing for International Market

Turning to the international market of PS and PG, in table 12.3 we find that in IC2,

1. The volume has increased from 7 percent to 14 percent.
2. The profit has increased from 8 to 14.9 units.
3. The share of profit has increased from 7.1 percent to 14.2 percent.
4. The ratio index of (PS/PG) units has risen from 40 to 180.
5. The ratio index of (PS/PG) profit % has risen from 61.4 to 195.8.

As the contribution of IC2 to the absolute volume and profits has increased, so has the importance of PS to the IC2 market. To cater to the IC2 market which finds the PS-type technology meeting its needs, it may be advisable to undertake licensing or joint venture efforts to *manufacture* PS items in IC2, in addition to marketing them. This is indicated in the technology transfer of PS from domestic to IC2 market with the directional arrow.

0.0 Shift to New Technology, PG—Manufacturing for International Market

Turning to IC1, we find that the volume of sales has increased from 9 percent to 40 percent, the profit units from 10 to 41.8, and the share of profits from 8.9 percent to 39.8 percent. With the rising importance of IC1 to the total market for PS and PG combined, we find that the PG products are able to meet the market's needs, the ratio indexes of PS/PG being less than 100. Considering the declining market in the domestic market segment, coupled with the increasing share of IC2, it may be well to introduce PG *manufacturing* to IC2, in addition to marketing. This is indicated in Stage C under "Manufacture" against the IC2 market segment.

0.0 Cars—CONCOLs Between the U.S. and Japan

In our discussion of *one* product, viz., cars, in the preceding section, from the point of view of the U.S., the "Domestic Market" will be the United States, and IC1, Industrial Country 1, could be Japan. Similarly, from the point of view of Japan, the "Domestic Market" will be Japan, and IC1, Industrial Country 1, could be the U.S. We can already identify CONCOLs in the manufacture of cars and their marketing in the following observation of the managing director of McKinsey & Co., in Tokyo and Osaka:

> None of Detroit's Big Three auto makers is purely "American." Ford Motor Co. has a 24.4% stake in Mazda and distributes its products in the U.S. Mazda in turn produces Ford products and distributes them through Autorama dealers in Japan. General Motors Corp. has a 5% stake in Suzuki Motor Co., and is negotiating with

> Toyota Motor Corp. to establish a joint venture in the
> U.S. GM is the largest shareholder of Izusu Motors Ltd.,
> with a 34.2% stake. (The No. 2 Isuzu shareholder has
> only a 4.8% stake).
>
> [Chrysler] has a 15% stake in Mitsubishi (an arrangement
> it has had since the 1960s) along with rights to distribute
> Mitsubishi products in the U.S. and it has a joint market-
> ing arrangement with Mitsubishi in Australia. . . .[2]

We saw in chapter 10 that the Fremont Plant of GM, which was closed down in 1982 would reopen under joint GM-Toyota ownership. Japan found that in the land-sparse homeland, it would cost $1.5 billion to duplicate the plant that GM could offer in Fremont. On its part, GM found that Japan would invest $1.5 billion in robotics to make possible the production of quality cars at cheaper price.

Even before the onset of the 5G revolution, the U.S. and Japan are pursuing CONCOL operations in trade and technology. The car manufacturers in both countries have mutual investment in each other's assets already, so that the segmentation of the future markets for both present products—improved—and potential products—innovative—made possible through 5G is but a logical sequel to current CONCOLs.

In our discussion of the *illustrative* manufacture and marketing of PS and PG cars, we found that in Stage C, joint ventures in PS would be appropriate for Industrial Country 2 and joint ventures in PG would be appropriate for Industrial Country 1. In other words, high-technology is segmented by markets, foreign (joint) production following exports. Ohmae points out that the U.S. operations in Japan generate more than $20 billion, compared to $5 billion produced by Japanese operations in the U.S.

0.0 Chips and Computers—CONCOLs Between the U.S. and Japan

The practice of manufacturing in one country for sales in the other is not confined to the U.S.-Japan relationship in cars:

> Even in the semiconductor industry there are many suc-
> cess stories: Texas Instruments produced $47 million in
> microchips for the Japanese market in the year ended
> March 31, 1982. (This figure doesn't include microchips
> it manufactured in Japan and shipped elsewhere.) In the
> same year, Intel expanded operations in Japan, Fair-
> child began work on a plant in Nagasaki, and Motorola

---

2. Kenichi Ohmae, "U.S.-Japan Trade Tensions Mask Close Private Ties," *The Asian Wall Street Journal*, 3 November 1983.

launched a new facility about 160 kilometers north of Tokyo. . . .

Hewlett Packard has a technical agreement with Hitachi
to produce 64K RAMs. Control Data Corp. has a cross-
licensing deal with Hitachi as well as a sales subsidiary
and supply tie-up with Takeda Riken Industry. . . .

Intel, which supplies the chip that is the electronic brain
of the IBM personal computer, also buys the same chips
from Hitachi and Oki Electric Industry. . . . And IBM recently formed a joint venture with Matsushita Electric
Industrial Co. to manufacture personal computers in
Japan.[3]

As in the case of cars, so also in the case of chips and computers, the ground
work is well-laid for CONCOLs between the U.S. and Japan.

0.0 Strategy of International Shifts in Technology and Trade—Multiple Products

Our illustrative discussion has been with respect to a single product. The 5G Revolution, however, will impact not only cars and computers, but a whole spectrum
of products and services now offered, and many that will emerge in the wake of
the revolution. The CONCOL arrangements between the U.S. and Japan on cars
and computers will be affected by the prospects for CONCOL in other products
from other countries. As we saw in Chapter 11, the fundamental forces underlying
the CONCOL arrangements of the U.S. and Japan are the changes in the industrial
structure itself. The shift in developed countries like Japan is from the basic steel,
basic petrochemicals, small appliances, ships and simple fibers to higher-valued,
more specialized segments of the same industries which bring higher gain to the
higher skills employed, and higher value-added to the industrial segments.

What are the options for the nonoil developing countries, such as Taiwan and
Korea, in the context of the shifting international industrial structure?

A drastic simplification of the past profile and present options are:

- 1965—low capital; low technology; low-skilled labor; simple products
- 1975—medium capital; medium technology; high-skilled labor; complex products
- 1985—high capital; high-technology; higher-skilled labor; more complex products?

Should the nonoil developing countries decide to make such a shift, they
would need to import high-technology from primarily three sources—the U.S.,

3. Kenichi Ohmae, *op. cit.*

Japan, and West Germany. It should be noted that the CONCOL criterion of playing with and against the same party or parties in the same game is fully satisfied in the high-technology imports today (from the U.S., Japan, West Germany) for high-technology product exports tomorrow (to the U.S., Japan, West Germany). The concomitant coalitions in trade and technology could be referred to as $(CT)^2$, for short.

$(CT)^2$ applies not only to the CONCOL relations between the nonoil developing countries and the high-tech sources, but also to CONCOL relations between the developed nations themselves. The G.M.-Toyota agreement to reopen the GM plant in Fremont is an example. The key to the agreement is the special *product* that the CONCOL would produce, viz., subcompacts. To meet that market demand, G.M. and Toyota find it advantageous to import Japanese robotics to the closed-down Fremont plant and manufacture something that would not otherwise have been manufactured.

It is quite conceivable that the subcompacts would represent an increasing degree of customization. With increasing affluence, the demand is for higher degrees of customization, the customers demanding that the car that one drives be unique because of special features. And this demand can indeed be fulfilled, thanks to computer programs that can make each unit in the assembly line different from the other:

> [Assistant plant manager of Nissan's Zama auto plant] Teuro Osawa explains: "The machine's sequence of movements is controlled by a central computer. Information is fed in telling the entire welding line they will be handling in order, say, a sedan, two hatchbacks, another sedan, a station wagon, and so on . . .
>
> [The flexible line assembly operation permits] The Zama plant, for example, [to produce] an average of 50 cars per month for the Danish market. In accordance with Danish law, each headlight incorporates a small "widescreen wiper," a feature not seen on any other of the thousands of vehicles pouring off the assembly line.[4]

The customized products are what we have termed performance-specific, PS. We could summarize the trade guidelines as follows:

In general, export PG products to nonoil developing countries, and export PS products to industrial countries.

4. Geoffrey Murray, "How Nissan Wows US Auto Execs," *The Christian Science Monitor*, 30 September 1982, 9.

0.0 Revolutionary Potentials of 5G Technology: Management

Of the three types of revolutionary potentials of 5G technology, the potentials in manufacturing are most readily seen, with the potentials in marketing less readily so, and the potentials in management the least readily so.

Yet, it is possible that the potentials in management of 5G technology could far outperform the potentials in manufacturing and marketing.

Management commits present resources to potential results. Since the results are not known in advance, better estimates of the results would indeed improve management. The results are produced, however, by interacting variables that are both inside and outside the system. Not all the relevant variables are usually identified; not all the relevant values of the variables are known; and not all the relevant interactions are known. The management question: How do I increase the ROI (rate of return on investment)? needs to be rephrased rigorously, the appropriate data have to be found, the proper methods of processing the data have to be identified, the correct analysis has to be carried out, and the valid conclusions drawn from the analysis, to make possible the best decision on the allocation of resources.

Dr. Tohru Moto-oka, chairman of the Committee for Study and Research on Fifth-Generation Computers, underscores the change in the Fifth-Generation Computer from numerical computations to, first, assessing the meaning of information, and, second, understanding the problem to be solved. In other words, faced with the foregoing management question, How do I increase the ROI? the 5G Computer should be able to assess the meaning of the question, understand the problem to be solved, and solve the problem using the appropriate databases. Moto-oka envisages four major roles for the 5G Computer:

> Fifth-generation computer systems are expected to fulfill four major roles: (1) enhancement of productivity in low-productivity areas, such as nonstandardized operations in smaller industries; (2) conservation of national resources and energy through optimal energy conversion; (3) establishment of medical, educational, and other kinds of support systems for solving complex social problems, such as transition to a society made up largely of the elderly; and (4) fostering of international cooperation through the machine translation of languages.[5]

Of the four, the second and third roles draw heavily upon the three-pronged capabilities of 5G, viz., intelligent interface, knowledge-base management, and

---

5. Tohru Moto-oka, "The Fifth Generation: A Quantum Jump in Friendliness," IEEE, *Spectrum*, 1983 November, 47.

problem-solving and inference functions mentioned earlier. Clearly well beyond the sophistication of Decision Support Systems and Expert Systems of today, the revolution in management promised by Fifth-Generation technology remains a consummation devoutly to be desired.

Appendix

Japan's Fifth Generation: A Progress Report

Asia On-Line, the Hightech business magazine, commissioned the author to assess the progress of Japan's Fifth Generation Computer Systems reported at the latest International Conference on the subject held in Tokyo, November 5–7, 1984. Of special significance is the personal discussion with the technical and administrative leadership in Japan, Europe and the United States, which is reflected in the three-part report published in the December, 1984, issue of *Asia On-Line*, whose kind permission to reprint the article is gratefully acknowledged.

The subject is covered in three parts: **Research Results,** *exploring some findings to date as well as questions that remain to be answered;* **Profit Potentials,** *looking at the market for products of 5G research; and* **American Assessment.**

Part I

Research Results

The melody, Momiji, familiar to Japanese school children, is played. Lyrics speak of red and yellow leaves reflecting autumn sunlight. The location is Tokyo. But this is not an elementary school, it's the Fifth Generation Computer Systems Conference. And the melody is played on the keyboard for a piece of metal to "hear." As several of the 300 overseas and 700 Japanese conferees witness, the Personal Sequential Inference Machine, or PSI, hears the melody and plays back a harmony.

What is so special about a composing computer?

The harmonizer itself is, as the Japanese say, a "toy" demonstration. But the research leading to it is nothing to toy with. It's just a small part of a large effort the Japanese are convinced will win them the world of 5G—Fifth Generation Computer Systems.

One way to appreciate the significance of the 5G Project is to look back at May 25th, 1961 when President John F. Kennedy said, "Land a man on the moon and return him safely to earth before this decade is out." It was a specific goal which could be recognized upon accomplishment, as indeed it was on July 20, 1969.

Similarly, the Japanese Ministry of International Trade and Industry (MITI) established an Institute for New Generation Computer Technology (ICOT) in

316

1982. ICOT's specific goal was to carry out a national project to develop a prototype 5G computer capable of serving as a knowledge-processing system by 1990.
Eight companies each contributed five researchers to ICOT: Fujitsu, Hitachi,
Matsushita, Mitsubishi, Nippon Electric Corp., Oki, Sharp and Toshiba. With 10
additional researchers, primarily from government laboratories, ICOT's research
team of 50 carries out the vital national project.

Differences Between 5G and Apollo
While the 5G is a highly visible national priority project with a prespecified, veri-
fiable end-product scheduled for delivery a decade hence as was Apollo, there are
several differences. Unlike Apollo, which essentially built a new capability out of
existing capabilities, 5G aims to create new capabilities to be incorporated into a
new machine.

Those new capabilities will constitute the foundation of the Information Society. Professor Tohru Moto-oka of Tokyo University, Chairman of the Committee
for Study and Research of Fifth-Generation Computers, identifies four major roles
for 5G computers.

"Fifth-Generation computer systems are expected to fulfill four major roles:
enhancement of productivity in low-productivity areas such as non-standardized
operations in smaller industries; conservation of national resources and energy
through optimal energy conversion; establishment of medical, educational and
other support systems for solving complex social problems, such as transition to a

George Chacko (right) and Tohru Moto-oka Chat at the ICOT conference.

society made up largely of the elderly; and finally, fostering international cooperation through the machine translation of languages."

Inference: The Key

What is knowledge processing?

When we enter the salary rates of individuals and ask a computer to print out individual checks, the computer multiplies the number of pay periods by the applicable rate. The input is the salary data, and so is the output. The data input produces data output, although in a different form.

Skipping intermediate steps, let us say that for a knowledge-processing machine, the input is requirements, e.g. sightseeing in Tokyo; and the output is knowledge, e.g. the sightseeing schedule which best fulfills requirements and resources. Requirements input produces knowledge output.

Three parts constitute the Knowledge Processing System or KPS:

1. An *intelligent interface* which enables spoken words in English, Japanese, or drawings and other symbols to be communicated by humans to machine;

2. A *knowledge base* which maintains and updates different sets of data required to answer specific problems; and

3. *Problem-solving* and *inference*, which use appropriate elements from the knowledge base to allocate resources to meet requirements, e.g. determine optimum sightseeing schedule; develop proper harmony to a given melody.

If pressed, the "K" in KPS means inference. Philosophically, knowledge is more than inference, which is the drawing of all valid conclusions under the given rules applied to the situation.

The notion of inference is the key to 5G. At the conference, the term was stoutly defended by Dr. Kazuhiro Fuchi, director of ICOT's research center. Fuchi calls the 5G computer an "inference machine."

"The conceptual framework of 5G," he said, "assumes the next generation computer as a predicate logic machine. We might call it an inference machine because the basic operation of predicate logic is inference. When we say inference machines, we mean basic inference such as is dealt with in symbolic language. We have developed sequential inference machines as tools for research in the intermediate stage 1985—1988 and beyond. One of them is PSI."

Inference and the Composing Computer

The young ICOT researcher who designed PSI and the harmonizer, Kazuo Taki, explained that musical notes are translated into numbers, e.g. do = 3, so = 2, by a new language called ESP. Two sets of rules are applied: progression and generation rules. By studying the entire melody, PSI determines the root note and, on the basis of the distance from the root note to others, the chord group—major or minor. Next, the first note is harmonized by applying rules of generation, then the second note. The two generated chords are passed through the tester. If rejected, harmonization is made until accepted by the tester.

*Sequential inference machine hardware - PSI (top), and
parallel type relational data base machine - Delta (bottom).
(Photos courtesy of ICOT)*

Inference is seen in the deduction of the progression and chords. The tester applies rules of consistency with and compatibility to the generated harmony and infers if the rules are satisfied or violated. The issue is not really music, but logic. The process is inference, not harmony.

PROLOG: The Future

Shakespeare wrote, "The past is prologue." But in 5G, we can say "The future is PROLOG."

To implement inference—if A leads to B, and B leads to C, then A leads to C—we currently use software. Programs tell the computer to carry out the operations specified. However, if inference rules could be built into hardware, automatic execution would be possible. The method by which programs are expressed in logical form and executed with inference functions is called logic programming.

Japan has chosen a language called PROLOG (**PRO**gramming in **LOG**ic) as the machine language of the logic processor. For logic programming to work, the knowledge base has to be arranged and rearranged according to different criteria as needed, a capability offered by relational databases.

Table A1
Research Accomplishments of Phase 1

The following are 16 accomplishments of Phase 1 of Japan's 5G research announced by ICOT at the November conference.

1. Inference subsystem.
   a. Investigation of various parallel inference schemes—data flow, reduction, clause unit, and complete copying scheme.
   b. Trial fabrication of hardware simulators.
   c. Accumulation and evaluation of design data.
2. Knowledge base subsystem.
   a. Development of a parallel type relation database machine
   b. Establishment of specifications for a tightly-coupled interface with the inference machine.
3. Basic software system.
   a. Establishment of parallel-type logic programming language (KLI) specifications.
   b. Development of large-scale specifications for knowledge programming language (Mandala).
   c. Development of large-scale relational database management program (KAISER).
   d. Partial prototyping of a knowledge utilization system (Japanese proofreader support system, etc.).
   e. Development of advanced syntactic analysis program (BUP).
   f. Trial fabrication of an experimental semantic analysis program.
3. g. Development of a modular programming system (for use with SIM).
   h. Trial fabrication of an experimental software verification management program.
4. Pilot models for software development.
   a. Development of sequential logic programming languages (KL0, ESP).
   b. Development of SIM hardware (PSI).
   c. Development of SIM software (SIMPOS).

Appendix

ICOT has announced 16 major results of its first phase of research, including the construction of the two machines—the PSI and the Delta, a relational database machine.

Part II

Profit Potentials
Ten years may seem a long way off, particularly as the target date for a prototype, as it suggests the product is even further down the road. Why worry about potential competition in the marketplace fifteen to twenty years down the road?

Gain the Whole World Market First
American businessman Frank Gibney multiplied Encyclopedia Britannica's Japanese business 17-fold to US$ 100 million in 10 years using Japanese methods. In his book *Miracle By Design*, Gibney characterized the Japanese approach to marketing as that of gaining the whold world (market) first. Edward Feigenbaum and Pamela McCorduck cite this success in their book *Fifth Generation: Artificial Intelligence and Japan's Computer Challenge to the World.*

"We now regret out complacency in other technologies," they write. "Who in the 1960s took seriously Japanese initiatives in small cars? Who in 1970 took seriously the Japanese national goal to become number one in consumer electronics in ten years? (Have you seen an American VTR that isn't Japanese on the inside?) In 1972, when the Japanese had yet to produce their first commercial microelectronics chip but announced plans in this vital 'made in America' technology, who would have thought in ten years they would have half of the world market for advanced memory chips?"

5G Applications
ICOT expects to apply 5G technology to a variety of markets. Formation of knowledge industries, conservation of energy and resources, and greater manufacturing productivity are just some of the expected applications.

To secure significant shares of yet-to-be-born markets, advances have to be pressed into practical applications quickly. While products are still in development, consumers-to-be must be convinced. While any imminent sale of products based on 5G breakthroughs is vigorously denied, the infrastructure to facilitate an effective market sweep is indeed well in place.

Phase 1 performance has been impressive. It is also clear that many more researchers than the original five from each of the eight corporations which contributed to ICOT have worked to achieve all that the 16 major accomplishments represent.

From one of those eight corporations, NEC, 32 conferees registered for the

Robotics/Artificial Intelligence/Productivity

Figure A4
Social Impact of 5G Computers

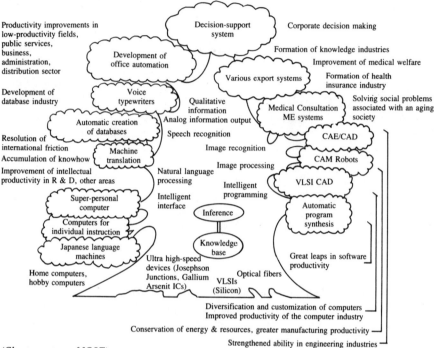

(Chart courtesy of ICOT)

conference. Allowing for some split attendance, such as attending the general sessions and various technical sessions, the actual number of participants and attendees may have been more like 35 to 40. Clearly, these people had direct and demanding interest in the 5G conference for NEC to allow them to attend. If NEC contributed five researchers directly to ICOT, and seven to eight times as many people are working on the 5G project at NEC, the size of the active network or researchers and developers on the project is not the 50 at ICOT, but probably more like 400.

MITI Incentives
What motivates Japanese industry?

Mr. Soichi Nagamatsu, deputy director of the Electronics Policy Division at MITI, explains that there are two kinds of programs to stimulate industrial technology: conditional loans and contract research.

MITI makes conditional loans to companies or consortiums to achieve a particular objective, such as strengthening the computer industry or one aspect of it

Appendix

George Chacko and MITI's Soichi Nagamatsu.

like, VLSI technology. The consortium research activity does not come under the purview of anti-trust regulations.

The loan is conditional in the sense that it is made on the condition that the loan be repaid when the strengthened technology brings profits. MITI settles in advance the time frame within which the technology will become profitable, say 10 years, and works out the repayment schedule at the time the loan is made.

The 5G project is not a conditional loan program, but a contract research program. MITI furnishes full funding for the ICOT project. All intellectual property rights accrue to MITI.

MITI will also determine the conditions under which research results will be commercialized, and will determine payback rates on the basis of potential profitability. In addition, MITI will issue patents to produce commercial products.

What do the eight companies get out of the 5G project?

The two machines produced in the first phase of the program were built by two teams of Japanese companies. Matsushita and Oki built the PSI, and Hitachi and Toshiba built Delta. The question occurs that this arrangement might give exceptional advantage to the companies building these embodiments of ICOT research.

According to MITI officials, the companies were selected on the basis of capability. They had built similar machines and had the knowhow, or could get up to speed fast. The engineering design is fully accessible to all of the ICOT family with ICOT permission. Mr. Nagamatsu did not find that every company was clamoring for access to every engineering design. The firms have several ongoing, full-time projects and so have little time or inclination to study every design.

Patent licensing to produce commercial products will be accomplished by let-

ting market forces operate. Companies will decide for themselves which particular market and product segmentation best suits them.

International Licensing

The eight participants in ICOT work together to advance the knowledge frontier cooperatively. But they are each taking aim at potential markets for potential products in which they will actively work against each other. This type of cooperative-competitive association is called Concomitant Coalition (CONCOL).

CONCOL applies to international trade as well. It is MITI policy to license patents to international users for equipment such as TVs and washing machines. When a country pays a patent fee, the country is working with Japan in furthering Japan's aim of higher reserves. However, when that country embodies the technology in products and exports them to markets currently served by Japan, that's competition. International licensing is thus a CONCOL.

In the long run, patent revenues can be as important, if not more, than product revenues. When Japan acquires 5G capabilities and moves on to acquire future generation capabilities, planners will concentrate on the most profitable segment of available licenses, selecting those yielding the highest returns, accordingly ranked 100, 99, 98, etc.

Licensing of Japanese technology with ranks of say 70, 69, 68, or even in the range of 80, 79, 78, etc., to Pacific region countries on a lower rung of the technology ladder could be doubly beneficial. It could provide Japan with a continuing source of patent licensing revenue and upgrade technology levels of the licensee.

It may well be that a long-range CONCOL strategy in international licensing of 5G technology can be developed to all around advantage.

Competitive Cooperation Essential

In his opening address, Prof. Moto-oka declared that competitive cooperation is essential in 5G computers. When asked if he meant CONCOL among the eight ICOT participants, he replied, "internationally." Moto-oka pointed out that 5G research sponsored by military agencies would probably not be made public, but in the case of unclassified research, nations can gain much by competing in some areas, but cooperating in others.

Part III

An American Assessment

The doubling of the number of nations represented, from 15 at the first 5G conference in 1981 to 31 at the second, which was held last month, demonstrates that the Japanese have attracted world attention to their assault on the unknown. The

conference, which was planned for 600, attracted more than 1,000 well exceeding the count in official lists.

A preliminary list of participants showed 875, with 518 or 59.2 percent from Japan, and 152 or 17.4 percent from the U.S. Nippon Electric Corp., one of the eight firms contributing researchers to set up ICOT, sent 32. This figure was closely matched by IBM, which sent 28 participants.

Participants from the Pacific region included eight from Taiwan, six from Korea, and one each from Singapore and Malaysia. There was no participants from the Philippines, Indonesia, Thailand, or New Zealand. Australia sent 10 and mainland China 18.

Among the Asian-Pacific nations, excluding Japan, Korea alone presented a paper.

The Decision to Dare
In analyzing essential elements of organizational success, I have coined the phrase, *"decision to dare."* This is the vision to venture beyond the familiar, committing significant resources to an uncertain outcome. Such a *decision to dare* was embodied in Kennedy's decision that the U.S. should commit what would be 10 times the total outlay of the largest previous national undertaking. Nobody ever having landed on the moon before, the decision that the U.S. would attempt to do so in the glare of world-wide attention was indeed a decision to dare.

Admiration of the Japanese decision to dare to invent the future in computers was universal among Americans at the conference.

"You've got to hand it to the Japanese," said one American conferee. "What they have achieved in this short time is impressive. They just picked up and applied whatever looked good. They made the right decision."

The primary decision has been "to reconstruct the body of computer hardware and software on the basis of a logic system called predicate logic," recalled Fuchi. He emphasized that PROLOG is not considered the final answer. Modifications are expected.

In designing a sequential inference machine, or PSI, an expanded version of PROLOG was designed as a machine language. We may look upon the machine language as the bips and bleeps the machine understands, translations of the user-language in which programs are written. Since the modified PROLOG was a machine language, a higher order language called ESP (Extended Self-contained PROLOG) was developed. Fuchi says that ESP is an answer to the initial hypothesis that "a predicate language is a promising, if not the only, candidate for a knowledge representation base."

Is choosing PROLOG a good decision? PROLOG, like anything else, has its good points and not so good points. Its logical calculus permits effective representation of knowledge, and it uses computationally fast methods. Detractors, however, say that knowledge represented in PROLOG is often hard to understand, and

the user is relieved of the knowledge of the problem-solving process, the very strength of AI or artificial intelligence methods.

AI—1990s Version

Fuchi emphasized that the 5G project was not spending time on present AI efforts because they use current-generation computers. Even if they are economically viable now, their limits will soon be reached. The Japanese want to concentrate on a 1990s version of AI.

At the 1981 conference, Prof. Edward Feigenbaum of Stanford University, a founding father of AI and one of the originators of the knowledge-based systems concept, advised the Japanese to pay special attention to expert systems. They took heed. At last month's conference, they revealed a knowledge representation system for a medical expert system designed to diagnose glaucoma.

What an expert system does is apply inference rules of experts, in this case a medical doctor. But given the same set of data the doctor sees, can the machine reach correct conclusions? The Japanese have developed three versions of the glaucoma diagnostic system. The object is to arrive at the diagnosis with the greatest probability. The machine retrieves the uppermost disease hypothesis and verifies it by applying all associated rules. If a lower level hypothesis emerges as tenable, its probability is calculated. The process is repeated until the uppermost disease hypothesis is established.

Three Priority Programs

The Japanese, with little experience in expert systems only two years ago, have proven they can select the right type of problem to learn, apply, and build on their knowledge. Of the three versions of the glaucoma diagnostic system, two used C-PROLOG and DEC-10 PROLOG.

LISP, the logic programming language most in use in the U.S., is not altogether abandoned. In fact, National Telephone and Telegraph is reportedly contracting for a high-speed LISP machine. Various Japanese papers reported work with LISP, and modifications of LISP. One such paper related a knowledge-based system for nuclear plant diagnosis. The approach applied both knowledge of normal and anomalous situations to diagnose trouble. The interesting thing, according to the paper, is that "the implementation was not in PROLOG, but in VAX 11 with Franzlisp to explore its diagnostic capability."

A national program in AI for nuclear plant diagnosis is apparently being given high-priority. VLSI design is another, and the SIMPOS operating system is a third.

In sum, as one expert put it, "I have to look very carefully at what they have done. Having worked on very large-scale computer projects, I know how quickly a demonstration machine like the PSI machine can be put up. But what's inside it? What functions can *really* be performed by it? They've studied the international literature very carefully, and are approaching the break-even point."

Index

Index

Dunne, M.J., 86n
Durant, Ariel, Will, 6n, 9–10, 9n, 10n

Edison, Thomas Alva, 203
employment, lifetime, 43, 51–3, 57, 219–25, 259
Engelman, Carl, 183n, 189n
Engelmore, Robert S., 182n, 189n
Engels, Friedrich, 10
engineering, x, xv, 43, 64–5, 81, 147, 149–50, 160, 232, 236
England(ish) (see also Britain(ish)), 3, 5, 7, 10–11, 13–5, 17, 19–20, 72, 124
entrepreneurs, see projectors
Erman, L., 191n, 192n
Ernst, H., 103n
expert(ise) (see also Expert Systems), 167–9, 182–3, *defn. 184*, 185–6, 188, 191n
Expert Systems (see also expert(ise)), x, xv, 148, 165, 167–70, 182, 182n, 183n, 184–6, *defn. 188*, 189, 189n, 315, 326

feeding, ix, xiv, 71, 83, 118
Fehling, M., 191n
Feignebaum, Edward, 154–5, 155n, 165, 165n, 183n, 188n, 189n, 193, 321, 326
fifth Generation Computer Systems (5G), ix–xi, xiii–iv, xvi–vii, 23, 136, 143, 143n, 148, 207, 264, 291, 293–4, 299, 301, 311–2, 314, 314n, 315–7, 320–4, 326
Fikes, R. E., 160, 160n, 161, 161n, 162n
Flexible Line Assembly (see also Flexible Manufacturing System (FMS)), 71, 96–8, 313
Flexible Manufacturing System (FMS)

(see also Flexible Line Assembly), 72, 76, 80, 95, 97–8, 149
Friedrich, O, 23n
Fu, K.S., 119n, 120n, 122, 165, 165n
Fuchi, Kazuhiro, 318, 325–6
Fujitsu, 55, 61
Funke, Michael, 240n

Garrison, R., 103n
Gaschnig, John, 182, 190n
General Motors (GM), x, xiii, xiiin, xvi, 17n, 121–2, 214, 246–7, 258, 300, 310–1, 313
Gibney, Frank, 41, 41n, 44, 44n, 45–6, 51, 51n, 54n, 56, 57n, 59n, 62, 62n, 63n, 218, 222, 223n, 227–8, 227n, 231, 257, 257n, 321
Gilriches, 38
gimmu 43, 48–9
giri, 43, 48, 220
Glushkov, Victor M., 158–9, 159n
Gonzales, R.C., 109n, 119n, 120n, 122–3
Goshi, Kohey, 24n
Gould, Rowland, 218n
Graves, S., 86, 86n
Griffith, R., 103n
Grossman, David G., 73, 74n, 103n
Gross National Product (GNP), GDP, 22, 28, 31, 38–9, 41, 59, 63, 77, 208–9, 241, 272
Group Technology (GT), 71, 91–2, 148
guidance, administrative (see administrative guidance)
Guo, T.H., 108n
Gusev, Leonid A., 158n

Haggerty, Patrick E., 125n
Haitani, Kangi, 56
Hamada, T., 120n
hardware, x, xv, 100–2, 123–4, 127, 136n, 139–40, 142, 148, 168, 191, 320, 325